Masters of Uncertainty

Masters of Uncertainty

Weather Forecasters and the Quest for Ground Truth

PHAEDRA DAIPHA

The University of Chicago Press   Chicago and London

PHAEDRA DAIPHA is assistant professor of sociology at Rutgers University.

The University of Chicago Press, Chicago 60637
The University of Chicago Press, Ltd., London
© 2015 by The University of Chicago
All rights reserved. Published 2015.
Printed in the United States of America

24 23 22 21 20 19 18 17 16 15   1 2 3 4 5

ISBN-13: 978-0-226-29854-2 (cloth)
ISBN-13: 978-0-226-29868-9 (paper)
ISBN-13: 978-0-226-29871-9 (e-book)
DOI: 10.7208/chicago/9780226298719.001.0001

Library of Congress Cataloging-in-Publication Data

Daipha, Phaedra, author.
  Masters of uncertainty : weather forecasters and the quest for ground truth / Phaedra Daipha.
      pages ; cm
  Includes bibliographical references and index.
  ISBN 978-0-226-29854-2 (cloth : alk. paper) — ISBN 978-0-226-29868-9 (pbk. : alk. paper) — ISBN 978-0-226-29871-9 (ebook) 1. Weather forecasting—United States. 2. Weather forecasting—Decision making. 3. United States. National Weather Service. I. Title.
  QC983.D26 2015
  551.63—dc23
                                                            2015005398

♾ This paper meets the requirements of ANSI/NISO Z39.48-1992 (Permanence of Paper).

To my parents,

who first and always nourished my quest for ground truth

Contents

	Introduction: Decision Making under Uncertainty	1
1	The Weather Prediction Enterprise	24
2	Working the Weather: A Shift in the Life of a Weather Forecaster	55
3	Distilling Complexity: Atmospheric Indeterminacy and the Culture of Disciplined Improvisation	92
4	Managing Risk: The Trials and Tribulations of Hazardous Weather Forecasting	112
5	Anticipating the Future: Temporal Regimes of Meteorological Decision Making	138
6	Whose Weather Is It Anyway? From the Production to the Consumption of Decisions	165
7	Toward a Sociology of Decision Making	197

Acknowledgments 219
Notes 223
References Cited 239
Index 261

INTRODUCTION

Decision Making under Uncertainty

Oh, what a blamed uncertain thing
This pesky weather is;
It blew and snew and then it thew,
And now, by jing, it's friz!
PHILANDER CHASE JOHNSON

Just another ordinary day at the National Weather Service, it seems. The two forecasters on shift are quietly fussing with their computers, deftly summoning animated weather maps in and out of existence as they click away at their screens. Just as they did the day before, and the day before that. Yet this is only my third time here and I cannot be sure of what I am observing. To make matters worse, I got caught in the thunderstorm on my way in and missed the 7:30 a.m. shift briefing. I tell Doug, the forecaster I am shadowing today, that I forgot to check the forecast last night, and he manages a smirk at the irony. But then he says, "It wouldn't have made much of a difference, actually. We, too, were caught a bit by surprise." Apparently, as late as last evening, the computer models showed no significant potential for precipitation. The following set of model runs early this morning, however, told a very different story. Doug has updated the forecast to reflect an 80 percent chance of showers and thunderstorms lasting through the midafternoon, diminishing to about 30 percent as the weather system exits the region. He is already working on the forecast of the next 36 hours but keeps darting glances at the radar loop and current weather observations pulsating on his left screen. Not that he would be able to stay focused on the forecasting task at hand, even if he wanted to. The surprise June thunderstorm, during a weekend, no less, has brought with it a barrage of phone calls from people concerned about their outdoor plans. Doug is on the phone with a woman who called about the forecast for the town twelve miles to the east. He pulls up the satellite imagery on his middle screen and then swivels his chair around to look out the window: "Well, things should be clearing up right about now, but there's still a 35 to 50 percent chance of showers and thunderstorms for our area." The woman must have asked for clarification because he goes on: "It's not going to be a washout, but there might be a few showers through the evening hours." At this point, as I learn later, the woman explains that her daughter is getting married at 5:00 p.m. and is it going to be raining then. Doug swivels back to his left screen to check the radar imagery once more: "Like I said, Ma'am, there's a chance of showers and thunderstorms this evening, although the radar doesn't show anything. So, conditions are a little unstable, but if you want a yes or no answer, I'd say no." The woman presses on. Doug, his voice rising a bit: "Ma'am. I cannot give you any guarantees. What I can give you is a 30 percent chance of showers and thunderstorms with a possibility of dense fog locally."

* * *

Weather forecasting is arguably the most iconic example of a decision-making task riddled with deep uncertainty—uncertainty that cannot be reduced to calculable risk. This irreducible uncertainty derives from the inherently chaotic nature of the atmosphere, popularly captured in the notion of the "butterfly effect." But it is further exacerbated by the relative sparseness of weather data, the imperfections of meteorological measurements, the dependency of numerical weather prediction models on unreliable initialization conditions, the competing needs of disparate forecast users. Hence the probabilistic language in which weather predictions are typically couched and that we conventionally like to attribute to forecasters hedging their bets. And hence the no less iconic jokes about these weather forecasting experts—for their perceived evasiveness and blunders in judgment but also for their hubris in claiming to be able to predict Mother Nature in the first place.

Contrary to popular wisdom, however, weather forecasters are not faring worse at mastering uncertainty than stockbrokers, physicians, professional poker players, or any other group of so-called expert decision makers. As a matter of fact, they are faring considerably better. They may not be, strictly speaking, masters of uncertainty—after all, by definition, no one can master deep uncertainty—but they certainly are masters at mastering uncertainty. Per the judgment and decision-making literature, weather forecasters exhibit an exceptionally high degree of reliability in their assessment of uncertainty and risk.[1] All the usual jokes notwithstanding, weather forecasts are continuously getting better. Not simply because computing power and numerical weather prediction models are getting better but because weather forecasters themselves are also getting better at assessing how good they are. What allows for such an effective calibration of their performance is the fact that, precisely because they are trying to predict the weather, they find themselves in the enviable position of being able to evaluate their decisions regularly and in near-real-time against ground truth information.

This rather uncommon property makes the weather forecasting task an exceptionally good case for theorizing the process of decision making. Following forecasters in their quest for the certainty of ground truth promises to hold the key, as it were, to unlocking the analytically elusive process of diagnosis and prognosis as it actually happens. And that, in a nutshell, is my objective in this book. By systematically excavating how weather forecasters achieve coherence in the face of uncertainty, how they harness diverse information to project themselves into the future, I endeavor to develop a conceptual framework for studying uncertainty management in action. I endeavor, in other words, to develop a sociology of decision making.

Theorizing Decision Making in Action

Our decision-making process is bounded by inescapable cognitive and informational constraints. Most of us, most of the time, make every effort to act "rationally," carefully weighing all available options to arrive at the best, most optimal decision. The more we strive to act rationally, however, the more it becomes apparent that we do not actually have all the possibly relevant information about the task at hand, that there is not enough time and resources to obtain all the possibly relevant information, that there are just too many possible future scenarios to consider even when there is enough time and resources. Indeed, the persistence of the pleonasm "most optimal" in common parlance serves as a testament to the fact that the quest after *the* optimal decision proves to be a chimera, impossible in practice. To further complicate matters, the process of decision making is often permeated with ambiguity: preferences are unclear or unstable, objectives shift, options come and go, preexisting solutions are strategically matched with problems rather than the other way around.

In practice, then, our decisions are not driven purely by choice. And they are not, and cannot be, perfectly rational. Letting go of the false notion that decision making is based on rational choice, however, does not require jettisoning rationality altogether, nor does it require resigning oneself to the idea that decision making is a fundamentally haphazard process. But it does require eschewing normative criteria in favor of context-dependent explanations of judgment in action. Once one acknowledges that decision making cannot aspire to some "gold standard" of optimality, then surely decisions are to be considered on their own terms, within the context of their production, and not according to some a priori definitions of rationality.

Back at the drawing board, any foray into the process of real-world decision making must begin by charting the contours of expertise in action. Given that, despite their best efforts, decision makers typically operate under varying levels of uncertainty due to cognitive and informational constraints, what does making a skillful if imperfect decision look like on the ground? A good place to start this conceptual charting expedition is with the paragon of skillful decision making itself: laboratory science. Indeed, the most celebrated contribution of science and technology studies, or STS as commonly known, has been precisely to crack open the black box of laboratory science and theorize the decision-making process inside. A long line of ethnographies of laboratory life demonstrate that, similar to the realities of production in "ordinary" workplaces, producing scientific knowledge is a messy, improvisational, and highly practical activity.[2] Rather than adhering to abstract laws and systematic rules of method, scientists creatively adapt resources at hand to solve problems as they arise. By reconceptualizing the process of scientific knowledge production as craftwork and scientists as pragmatic entrepreneurs aiming for success rather than truth, STS scholars opened the door for the reconceptualization of expert decision making more broadly. Expertise is now seen as predicated on shared, embodied skills and know-how arising in situated conditions of use. Becoming an expert presupposes socialization and apprenticeship into the meaningful conventions and techniques of the community of experts one wishes to join—the acquisition of a discipline, in effect. But to be recognized as an expert is a never-ending process requiring relational work and a continuous makeshift readjustment of one's knowledge claims relative to those of other experts. Expertise must be enacted and performed to be socially salient, to exist in practice.

This is how I employ the term *expertise* in this book. If one is to eschew normative criteria of human rationality in search for empirically grounded accounts of judgment in action, then one must remain equally wary of efforts to derive a normative theory of expertise (cf. Collins and Evans 2006). And so, while I note the various skills and competences that are marshalled in the ever-present quest to master uncertainty, I regard expertise as a relational achievement, provisionally negotiated among humans, machines, and nature within a particular sociotechnical context of practice. By extension, my empirical focus on "expert" decision making is not meant to highlight an ideal or an exemplar but simply to take advantage of the fact that the process of decision making is as a rule considerably more externalized, hence considerably more observable, in settings where "experts" are at work.

Charted out of what we already know about expertise in action, my approach to studying decision making in action is guided by the following core assumptions. (1) Decision making takes place within a more or less institutionalized field of action that, over time, affords its members a certain stock of knowledge. (2) This stock of knowledge consists of cognitive heuristics and decision-making techniques that help initially frame and specify the empirical context of action. (3) Decision-making action may not be "rational," but rarely is it routinized or unreflective—instead, it is habitual and eminently practical. (4) It is within the evolving microcontext of action and the human and nonhuman others populating it that decision making takes form first and foremost.

I progressively flesh out and expand on these principles underpinning my theory of decision making in action through the case of weather forecasting. And, in the final chapter of the book, I pull them together into a conceptual framework for the decision-making process that I elaborate further through the cases of medicine and finance. For now, let me briefly outline how each of these distinct but interrelated principles contributes to a more robust understanding of how decision making actually happens.

Decision Makers Operate in an Institutionalized Environment

No study of the decision-making process can afford to ignore the role of the environment in which decision makers operate. Seventy years ago, Herbert Simon, James March, and their collaborators debunked the myth of the "rational man" and paved the way for more rigorous research on the "bounded rationality" of real-world decision making precisely by drawing attention to the environmental as well as the cognitive constraints structuring human behavior.[3] Even as it acknowledges the inherently bounded rationality of all decision makers, however, the bulk of the literature on judgment and decision making in cognitive psychology and behavioral economics today continues to cling to the optimality chimera to the detriment of real-world evidence. The fact that decision makers pursue "satisficing" rather than "maximizing" solutions as a matter of practice (Simon 1957, 204) has effectively been rendered analytically irrelevant. Decision making is divorced from its natural environment, and attention is instead devoted either to identifying suboptimal types of judgment via laboratory experiments and simulations or to devising optimizing models of bounded rationality.[4]

Yet it bears underscoring that only after shifting the unit of analysis from the individual to the group—or, most famously, to the organization—and theorizing the process of decision making in specific fields of action was Simon able to formulate his theory of bounded rationality. The moment it is examined as a social phenomenon, nonreducible to individual cognition, the process of decision making in the real world, although still imperfect, emerges as rational once again. For, in the course of enacting expected roles, decision makers come to rely on, and identify with, in-group conventions and performance standards, and they acquire an "organizational personality": they turn into an *"organized* individual" (Simon 1947, 278, emphasis added). Simon's reconceptualization of the "rational man" into an "administrative man" is not without its problems, as I discuss below, but it does provide an important opening for studying rationality as an externalized cognitive structure. Fifty years later, distributed cognition theory extends the scope of the decision-making environment from the social to the sociotechnical realm by drawing attention to the material culture and procedures enlisted to help store, process, and coordinate information during complex decision-making tasks (Hutchins 1995a, 1995b).

Within sociology, meanwhile, the notion of the organization as a cognitive system has been fruitfully combined with field theory (cf. Martin 2003) to model the emergence and institutionalization of decision-making environments at the interorganizational and supraorganizational level (DiMaggio and Powell 1983; Powell, Packalen, and Whittington 2012). More recently, macroanalytic focus on the cognitively constraining influence of institutions is increasingly being complemented by attention to the microfoundations of organizational behavior. What drives efforts to bridge the micro-macro divide is growing recognition of the internal tensions and contradictions inherent in the logic of any institutional field given the complex of cultural symbols and material practices through which decision makers understand and organize their world (Thornton et al. 2012, 2; Friedland and Alford 1991, 248). Emphasizing that institutional contradictions provide both opportunities and constraints for creative local action, scholarship on institutional logics has generated new insight into how organizational actors transform and reproduce the environment in which they operate. Yet, having drawn attention to the importance of material practices, it typically stops short of actually attending to the host of artifacts and technologies through which logics become articulated. As a result, it has been at pains to account for the emergence and institutionalization of organizational routines without ultimately resorting to some kind of a top-down deus ex machina explanation of social order. There is increasing realization in the literature that a greater attention to the interlocking, or "imbrication" (Leonardi 2011), of the technological and the organizational "could contribute to a richer understanding of the constitution, constraints, and affordances of institutional logics" (Cloutier and Langley 2013, 364).

Decision Makers' Stock of Knowledge Frames the Empirical Context of Action

Once one commits to considering decisions on their own terms, within their proper environment, it will not do to then proceed to study them removed from the actual scene of their production. Granted, it is no easy task to upend the original scholarly fixation on decision makers as rational, utility-maximizing actors but still maintain an appreciation for human actors as decision *makers*. Simon's "administrative man," despite being a significant improvement over the previously prevailing decision-making model of the "rational man," has been variously criticized for offering a no less unrealistic portrayal of how decision makers operate in organizations. In the words of Langley et al. (1995, 266–67; emphasis in the original), Simon's decision makers are "perhaps more life-*size* [but] hardly more life-*like*." Yet, if we are to take decisions on their own terms, we must reserve the same consideration for the people who actually produce them. In practice, however organized they may be, however much they may rely on institutionally sanctioned rules and procedures to efficiently carry out a task, decision makers can typically not afford to blindly follow protocol but must remain tactically alert to the task at hand, creatively adapting rules and procedures to the evolving circumstances of action and interaction. Far from being passive executors of the logic of the institutionalized environment in which they operate, actual decision makers are hard at work to successfully enact this logic into practical solutions, to make it workable, routinely exercising considerable discretion in the process. The gritty realities of decision making as situationally driven improvisation and craftwork, however, are consistently obscured in most accounts of organizational life. Explicitly or implicitly, decision-making action is treated as just another "organizational output" (Allison and Zelikow 1999, 168ff.), with decision makers adhering mindlessly and inexorably to standard operating procedures.[5] Simon's administrative man lives on.

In no small part, the lack of sustained attention to the decision-making process as action in the extant literature (Nutt and Wilson 2010) can be directly attributed to the lack of a critical mass of sociologists expressly committed to the study of decision making as a sociological topic in its own right. The field continues to be dominated by cognitive psychologists and behavioral economists, while sociological contributions, although significant, remain scattered across a variety of substantive and analytic foci without a shared theoretical scope. Perhaps the most prominent of these has been scholarship that accounts for organizational disasters and failures.[6] Although scarce, disaster studies make critical inroads toward a sociology of decision making because—above and beyond the usual sociological preoccupation with issues of organizing, group culture, innovation, or change—they are explicitly concerned with unearthing the reasons that led to a fateful decision. Adopting a systems approach to decision making, such studies are attentive to the "latent pathogens" (Reason 1990)—such as insufficient training, deferred maintenance, substandard communication, or dormant design flaws—that can contribute, often in cumulative ways, to a poor decision. Hence, poor decisions are no longer conceptualized as the result of faulty reasoning. And the dark side of organizations does not reside in the cover-up of mistakes and misconduct but, more profoundly, in the gradual routinization of nonconformity and, therefore, in the perilous expansion of what counts as acceptable risk (Vaughan 1999a, 1999b).

Still, because they are primarily interested in explaining decision failures, disaster studies arrive at the scene after the fact—too late to truly capture decision making in action.[7] As a result, they tend to analytically privilege the logic of the organization over the individuals and groups embedded in it, thus undertheorizing the space for creative, strategic action afforded in practice by the various sociotechnical tensions inherent in any decision-making field. Needed are accounts of decision making that simultaneously illuminate both the institutionalized environment in which actors operate and the empirical context in which decision making actually unfolds. A notable exception is the work of organizational psychologist Karl Weick (1993; 1995), whose influential theory of *organizational sensemaking* draws attention to the retrospective rhetorical rationalizations decision makers frequently engage in to invent order out of informational equivocality as a result of the loose coupling between institutional logics and the empirical microcontext of decision-making action. Yet, whereas sense making—and risk management more generally—is often retrospective, the process of decision making is fundamentally *prospective*.[8] Capturing how we make decisions calls for research designs that are both action centered and forward looking.

The recent turn to naturalistic studies of decision making in behavior economics and cognitive psychology marks an important step forward in this respect, but it has yet to systematically consider the cultural and broader social parameters of organizational life.[9] Within sociology, meanwhile, although focus on the process of decision making in its own right is lacking, ethnographies of expert organizations have made great strides in bringing "individuals back into institutional theory" (Lawrence, Suddaby, and Leca 2009, 52) by shedding light on the local embeddedness of organizational meaning making and demonstrating that institutions are inhabited by individuals and groups doing things together (Hallett and Ventresca 2006, 213; Fine and Hallett 2014, 29). Despite occasional nods toward the role of material culture in shaping organizational life and group practices, however, they remain inattentive to the dynamics of human-nonhuman relations, their analytic gaze exclusively fastened on the realm of symbolic interactions.

To coax a more realistic model of decision-making action, I conceptualize the role of the institutionalized environment as primarily geared toward affording decision makers with what Alfred Schütz (1970, 74ff.) has called a "stock of knowledge": a shared, established repertoire of techniques, rules of thumb, and other decision-making habits meant to make the decision-making task tractable in appropriate, institutionally sanctioned ways. Schütz's stock of knowledge nicely encapsulates how organizations seek to steer individuals and groups toward rationally efficient decisions (Simon 1947). But it can also be extended to account for the dual nature of rules as both prescriptive and performative structures. It thus underscores the sociomaterially improvisational and context-specific character of decision-making practice by elaborating how actors make do with, reconfigure, and provisionally transform available tools and information to confront the inevitable messiness of concrete decision-making situations. Hence, decision makers are once again reinstituted as the *makers* of decisions, creating locally rational solutions out of the more or less institutionalized heuristics and techniques with which they have equipped themselves over time. The unit of analysis now is neither the individual nor the organization, but the task at hand.

INTRODUCTION

Decision Making Is Habitual and Practical Action

Perhaps the most fundamental premise of my approach to decision making is that reasoning is always inexorably practical. Experts and expertise are established within particular "communities of thought" (Fleck 1979) precisely because these are also "communities of practice" (Lave and Wenger 1990; Wenger 1998). Action properly constitutes and organizes judgment into being rather than simply turning it into an observable object of study (cf. Suchman 1987). My analytic foregrounding of decision making as externalized practice, therefore, very much represents a theoretical choice and not a methodological by-product. As such, this book is in step with the ongoing "turn to practice" across the social sciences (Schatzki, Knorr Cetina, and von Savigny 2001). Yet, whereas theories of practice typically emphasize the routinized, unconscious, and unthought aspects of action, my focus on the practice of decision making—a seeming oxymoron—also allows, indeed strains, for its conscious, deliberative, and meaning-making aspects. Surely making a decision is often not a mere matter of deploying a skill or passively carrying out institutionally sanctioned protocols but entails a creative response—both during nonroutine and during routine situations.

That decision-making practice is not as routinized and programmed as most conceptualizations of practice and decision making would suggest becomes readily apparent in knowledge-based settings, where practitioners are expected as a matter of course to continually keep streamlining and improving how they acquire and implement information. While hardly the only case in point, scientific practice, once again, offers the most obvious example of a decision-making activity that can be both habitual and not routinized. Hence the appeal of STS scholarship when one is in search of leads on how to keep analytically attuned to the whole range of decision-making practice. One runs across multiple, disparate, even conflicting approaches to accounting for scientific practice in the extant literature. But they all famously share one element in common: a rejection of the all too familiar human-centric bias of conventional theories of practice in favor of a properly object-oriented epistemology. Bruno Latour (1992) thus calls on us to rethink our disciplinary assumptions and allot equal consideration to the nonhuman "missing masses" that make up the fabric of social life. And Karin Knorr Cetina (2001) proposes a relational theory of "objectual practice" that conceives of knowledge work as fundamentally constituted in and through interactions with material objects rather than just other social actors. In the same vein, Trevor Pinch (2008) makes a case for integrating institutional theories of organizational behavior with a sociology of technology. Regardless of their specifics, all such conceptualizations point to the key role of the materiality—not just the corporeality—of action in structuring day-to-day problem solving. Yet, even in management studies, which have thus far been leading the effort to link organization theory with STS insights on sociomateriality, the overwhelming bulk of the literature continues to not take into account the role of technology in organizational life (Orlikowski and Scott 2008).

The notion that action is constituted in and through material relations has deep theoretical roots—most explicitly in the philosophical tradition of American pragmatism. Etymologically derived from the Greek *pragma*, or "thing," which shares the same root with such English words as *practice* or *practical* (cf. James 1907), *pragmatism* is founded on the materiality of practice and experience. "The whole originality of pragmatism," writes William James (1909, 115-16), "the whole point of it, is its concrete way of seeing. It begins with concreteness, and returns and ends with it." Pragmatism has strongly influenced constructivist and situationist approaches in contemporary sociology, but this influence has been mostly indirect, mainly attributed through pragmatism's influence on symbolic interactionism. Yet what gets lost by the lumping together of pragmatist and symbolic interactionist theories is precisely the materialist grounding of experience, which remains ignored or at best implied in most sociological accounts of action and interaction.[10] For pragmatists, as John Dewey (1922, 18) has so pointedly articulated, experience is "double-barrelled": "It recognizes in its primary integrity no division between act and material, subject and object, but contains them both in an unanalyzed totality. "Thing" and "thought" . . . refer to products discriminated by reflection out of primary experience."

Of late, sociologists are turning to pragmatism in increasing numbers.[11] Although the materiality of action continues to be left in the background, what does get foregrounded by this "pragmatist turn" in sociology is the practicality of action. Action at last emerges as decidedly nonteleological and nonnormative, with social actors motivated by ad hoc real-world problems rather than preset goals, values, or beliefs (Joas 1996; Swidler 2001, 81ff.). As such, while purposive, action is inherently flexible, improvisational, and opportunistic. Problem solving—and decision making more generally—hardly adheres to a premeditated and internally coherent course of action in practice, despite one's best intentions or avowals to the contrary. Instead, actors are confronted with a messy ongoing flow of situations, where ends blend into means and it is the resources at hand that lend concrete shape to goals in the first place (Dewey 1939, 40ff.; Whitford

2002, 37–38). Intense awareness eventually gives way to habit, of course, and most action ends up following a well-tried and familiar path until it stumbles on what seems a novel or challenging situation. Even habitual action, however, need not be mechanical and mindless but often constitutes a quite intelligent performance, betraying the thoughtful, nuanced sensitivity of an experienced practitioner. John Dewey's (1922, 65–66) crucial but overlooked distinction between intelligent habit and routine habit bears quoting at length:

> All habit involves mechanization. . . . But mechanization is not of necessity *all* there is to habit. . . . Mechanism is indispensable. If each act has to be consciously searched for at the moment and intentionally performed, execution is painful and the product is clumsy and halting. Nevertheless the difference between the artist and the mere technician is unmistakable. The artist is a masterful technician. The technique or mechanism is infused with thought and feeling. The "mechanical" performer permits the mechanism to dictate the performance. It is absurd to say that the latter exhibits habit and the former not. We are confronted with two kinds of habit, intelligent and routine. All life has its élan, but only the prevalence of dead habits deflects life into mere élan.

To say, then, that the practice of decision making is habitual is not to say that it is routinized. On the contrary, within a pragmatist perspective the structure/agency binary becomes relegated to the two extremes of decision-making practice.[12] Unlike Bourdieu's top-down explanation of the process of habituation, Dewey regards the embodiment of habits as an ongoing intelligent "achievement" (Dewey 1922, 38ff.). For Dewey, as John Levi Martin (2011, 263) puts it, habit stands "for our general capacity to "tool ourselves" to fit the world" (see also Hickman 1990). Decision making is discerningly creative and skillfully resourceful precisely because it is based on practiced, habitual ways of reasoning and doing. Ann Swidler's influential reconceptualization of culture as a "tool kit" of incongruous—because pragmatically assembled—resources evocatively captures the sort of situational improvisation that decision makers engage in as a matter of course. Swidler (2001) focuses primarily on the semiotic dimension of cultural repertoires, but her conceptualization can be readily extended to accommodate a more materialist understanding of culture, where meanings are proliferated through "epistemic objects" (Knorr Cetina 2001), articulated into blueprints for action, and rendered intelligible through bodily experience (see also Lizardo and Strand 2010). Decision-making habits are rarely passive or frozen because decision makers are charged with *making* decisions—with combining available resources to fashion a provisionally coherent solution to the routine and nonroutine challenges of the evolving task at hand.

Because it actively avoids a priori assumptions about the nature of practice and because it invites a symmetric consideration of both the social and the material conditions of action, I enlist pragmatist theory to serve as the basis for my theory of decision making. With it as the backdrop, I am able to draw out and meld together the complementary strengths of STS scholarship on knowledge production with cognitive, behaviorist, and sociological elaborations of bounded rationality. Thus refocused, the STS effort to debunk mentalistic explanations of expertise by establishing that reasoning and rationality take form outside the brain, within sociotechnical ecologies of practice, becomes a natural extension of pragmatist theory of action. And, translated into a cognate idiom, studies of organizational decision making are brought into conversation with studies of laboratory science: the former introduce an all-important institutional dimension to knowledge production, while the latter insist that decision makers are not "judgmental dopes" (Garfinkel 1967, 67) but purposeful actors creatively negotiating the microcontingencies of their environment.

Decision Making Takes Form within the Microcontext of Action

The last guiding assumption about the process of decision making follows naturally out of the three previous ones. I will therefore refrain from additional programmatic statements and wait instead until subsequent, empirically grounded chapters to make the case that the unit of analysis best suited for capturing decision making in action is the task at hand. It is this last, and most crucial, guiding assumption that allows for the socialized, processual, materialist, and dynamic character of decision-making practice to fully emerge into view. To showcase the role of the empirical context in shaping meteorological decisions, I have endeavored as much as possible to present decision-making action in the form of entire episodes rather than snippets. This is particularly the case in the later chapters of the book, once readers have become familiar with basic aspects of weather forecasting operations. In doing so, I am able to illuminate the reciprocal relationship between processes of prognosis and diagnosis in the making of decisions. Contra conventional decision-making models, which take decision ends as fixed and only inquire into the means used to attain them, my theory of decision making relies on Dewey's conceptualization of the dynamics between prognosis and diagnosis as an ends-means continuum: faced with a challenging situation, decision makers entertain several possible courses of action, which are simultaneously preliminary descriptions of the problem and its solution; these incomplete descriptions prompt the gathering of new information and renewed, iterative prognostic evaluations of the best course of action until a decision is reached—that is, until a satisfactory *joint* description of the problem and its solution has been achieved. Truly, as I shall show, decision making takes form within the microcontext of action.

Having settled on the appropriate analytic lens for studying decision making, it is now time to settle on the appropriateness of my chosen case study. This is a common enough exercise, of course, expected of all scholarly research. But, given the scope of this book, I have assumed a higher burden of proof, as it were. For it will not quite suffice to claim that weather forecasting constitutes a good site for studying decision making. I must perforce make the case that it constitutes an exceptionally good site—a "theoretically strategic" site, to use Jack Katz's (1997, 412-14) formulation. In the following section, therefore, before launching into a description of my data collection strategy, I outline the reasons forecasting operations at the National Weather Service offer a rare opportunity to study in a richly concentrated form the typically elusive process of how we make decisions.

The Case for/of Weather Forecasting

Whatever one may think of the quality of weather forecasts, the link between weather forecasting and decision making should be rather obvious. To be a weather forecaster means to live in what Robin Wagner-Pacifici (2000, 3) describes as a "subjunctive mood": a dispositional openness to possibility and contingency due to the fundamental indeterminacy of the decision-making context. Already, the deep uncertainty of the weather has brought much scholarly attention to the experts charged with predicting its behavior. Cognitive task analysis in particular—a relatively recent methodological approach to the study of expertise that marries behavioral economics with cognitive science and industrial psychology—has found much appeal in weather forecasting practice as an expert system of uncertainty management. More to the point, it has helped highlight weather forecasting as action in the subjunctive mood. Efforts to elicit and model the cognitive processes guiding the forecasting task underscore the absence of one single optimal method for mastering meteorological uncertainty and, instead, trace the multiplicity of reasoning styles and decision-making strategies forecasters adopt as a response to the variable weather regimes and information resources they must contend with.[13]

To be sure, consistent with other naturalistic behavioral economic and cognitive research on decision making, studies of weather forecasting have a blind spot for weather forecasting's social and cultural context. Even so, however, one can still easily glean from those studies that if forecasters have mastered how to operate in the contingency of the subjunctive mood, this is because they have become inculcated into an "aesthetic of imperfection" (cf. Weick 1999) that gives them license to make pronouncements that are doomed to always be provisional and incomplete, forever dependent as they are on the latest reports of how the weather is materializing and soon made irrelevant by new atmospheric developments. With its *culture of disciplined improvisation*, as I will show, the case of weather forecasting is thus particularly well suited to capturing the routine and nonroutine aspects of decision-making practice.

But further, the markedly prognostic emphasis of weather forecasting operations offers an excellent vantage point from which to observe the ends- or future-oriented character of the decision-making task while also attending to its practical, presentist logic. What is more, this vantage point displays the process of uncertainty management against a variety of time horizons such that distinct regimes of decision making emerge into view. Not only does the indissoluble link between meteorological ends and means become apparent, then, but one is also in a position to trace how prognosis merges into diagnosis and vice versa in the course of a given shift and how this coconstitution plays out across different temporalities of decision making.

Even though weather forecasting is obviously well suited to a case study of decision making, it is perhaps less obvious that the National Weather Service (NWS) would be the right empirical setting for conducting such a study. The NWS, to be sure, is the government agency charged with providing meteorological warnings and forecasts for the United States. Overwhelmingly, however, the weather forecasts consumed by the American public today are produced by the private sector and broadcast media (Harris Interactive 2007; Pew Research Center 2011). One may be inclined to counter that these forecasts are simply cannibalized and repackaged versions of the NWS forecast, but this is hardly the case anymore: over the past three decades, a fast-growing private weather industry has been creating predictions that rival in skill those of the NWS.

Still, while it is true that the NWS no longer holds a monopoly on authoritatively forecasting the weather, what makes NWS forecasting such an attractive case for the purposes of theorizing decision making is that it best allows insight into the material and institutional dimensions of the forecasting task. After all, it is because of the "knowledge infrastructure" developed and operated by the NWS and other government weather agencies around the world that weather forecasting as we know it today is at all possible (Edwards 2006). Without this day-to-day coordination of weather stations, radar, sounding balloons, and numerical weather prediction models, it would be impossible for government and commercial meteorologists alike to produce the weather forecast, and the current process of forecasting the weather would be unthinkable in itself. Meteorological or otherwise, reasoning originates in practical activity, not some abstract ideational realm. "All deliberation is a search for a way to act," writes John Dewey (1922, 119); "Deliberation has its beginning in troubled activity and its conclusion in a course of action, which straightens it out." Because it gets enlisted to solve concrete problems, therefore, reasoning does not reside in the mind but in its interface with a host of artifacts that are specifically designed to assist decision makers in securing and processing otherwise humanly inaccessible information. That is precisely why we have not been content with just being experts at interpreting weather signs, Dewey says elsewhere (1910, 16), but have instituted a weather service equipped with special instruments and techniques for detecting weather signs in advance of their appearance. This way, even if we cannot ward off the weather altogether, we may at least protect ourselves from its full effect.

For its key institutional role in things meteorological in the United States and for its monumental externalization of the forecasting process into observable techniques, procedures, and documents, I thus have sought to gain entry into decision making in action via an ethnographic study of forecasting operations at the NWS. Note that contrary to popular perception, NWS forecasting operations are not housed in a centralized meteorological facility but in 122 forecast offices across the country, assigned to and located in a specific geopolitical region. In effect, what is commonly understood as the NWS forecast in reality involves one out of more than a hundred separately issued and disseminated NWS forecasts, each catering to a particular locale but digitally stitched together into an apparently seamless national whole. The bulk of the action that informs these pages, therefore, takes place on the operations floor of one such local NWS forecast office in the northeastern United States, an office I call "Neborough."[14]

Fieldwork at the Neborough office occurred over the course of five years: fourteen months from 2003 through 2004 and another eight months in 2008. After an initial intensive three-month familiarization period in which I spent five days a week at the office, I settled into a four days a week schedule, which turned into weekly three-day visits when I returned to the field in 2008. In order to follow the weather and gain access to the quotidian minutiae that set the tone and illuminate meteorological decision making in action, I would arrive a half hour before and remain a half hour after the forecasting shift I was to observe that day, typically staying at the office for at least eleven hours during each visit. The only exception to the above observation schedule was the eight separate times I shadowed the graveyard shift. Just like forecasters, I clocked in for seven nights straight at a time in an effort to, as closely as possible, bodily engage (cf. Okely 2007) with the physicality of NWS forecasting. Just like forecasters, furthermore, I would stay longer or, on rare occasions, overnight at the office during major hazardous weather events.

As grueling and intense as this fieldwork regimen was, it was also necessary. Above and beyond the obvious substantive gains, it helped me quickly become conversant in the principles of weather forecasting and the operational logic of the NWS. Although I have tried to stay clear of meteorological jargon and overly technical explanations in this book, they were crucial for me to reliably and convincingly participate in the action on the Neborough operations floor. In the process of soaking up all I could about NWS forecasting, I inevitably caught the "weather bug" and developed a genuine fascination for the science of meteorology. Yet, like most every ethnographer eager to become one with the natives, despite my keen and conscientious study of meteorology, I can at best lay claim to an "interactional expertise" (Collins and Evans 2002): I can follow the process of weather forecasting and relay back what transpired to the satisfaction of a seasoned forecaster, but I would never presume to be capable of actually producing the NWS forecast itself.

As well, there can be no doubt that this punishing regimen quickly earned me the respect and approval of the Neborough forecasting staff and legitimized my position in the office. Within three months' time I had been adopted as the "in-house sociologist." And while I never became one of the guys, I will forever be grateful to the guys for treating me like one of them. In due course, I was not only made part of office baseball outings and a retirement party, but I was also invited to informal restaurant gatherings, people's homes, two weddings, and so on. More to the point, with the exception of labor union meetings, I was granted unlimited access to any and all facets of Neborough forecasting operations, including relevant paperwork, chat room discussions, office meetings, training workshops, and outreach events. Finally, twice as my time at Neborough drew to a close, I conducted in-depth interviews with all but one member of the entire staff. All interviews were audiotaped, in contrast to my field notes, which were primarily taken on site using shorthand. Quotes and dialogue throughout the book are to be taken literally, therefore, and are not reconstructed from field observations.

With its focus on the everyday microprocesses underlying meteorological decision making, this book follows in the tradition of a long line of shop-floor and laboratory ethnographies that conceptualize knowledge production as craftwork. Although grounded in the particular sociomaterial context of forecasting at the Neborough office, its objective is to capture prevailing patterns of meteorological decision making across NWS offices and indeed across the entire field of weather forecasting practice in order to generate a sociological theory of decision making in action. This mesolevel perspective I have endeavored to attain through a fourfold strategy. First, I identify the organizational logic behind NWS forecasting practice by pursuing an institutional account of the recent controversial NWS shift from a text to a graphical forecast. The study of controversies has proven an important methodological tool for making visible the normally hidden processes that sustain and regulate systems of expertise (for a review, see Pinch and Leuenberger 2006). By tracing the unfolding and eventual closure of this operational controversy, which I was fortunate enough to observe on the ground, I am thus able to probe deeper into the cultural, infrastructural, institutional, but also broader social parameters that undergird weather forecasting operations at the NWS. Second, I leverage Gary Alan Fine's (2007) comparative ethnography of three NWS offices in the midwestern United States to draw broader conclusions about NWS forecasting practice beyond the specifics of the Neborough case. Although not systematically concerned with the process of meteorological decision making, Fine's superb analysis of the culture of NWS "futurework" nonetheless provides a much-needed point of confirmation, and occasionally juxtaposition, for my argument. Third, I follow meteorological decision making up the vertical structure of the NWS. In June 2004, I spent a day at NWS Eastern Region Headquarters, where, in addition to conducting interviews with the director, deputy director, and seven department heads, I sat in on the daily regional weather briefing and a senior staff meeting. This fleeting bird's-eye view on NWS eastern region field operations helped me ascertain to what extent what I was observing at Neborough was typical across all twenty-three offices of the region, and it solidified my impressions about the administrative perspective regarding the shift to a graphical forecast. Fourth, I follow meteorological decision making across the weather industry. Between May and July of 2004, I interviewed eight broadcast meteorologists stationed in the biggest media market of the Neborough region. While I did not get an opportunity to also spend time at a private weather company, I was able to conduct short interviews with two forecasters from Meteorlogix during their visit at the Neborough office in 2004. I was thus able to establish what is and is not specific to NWS forecasting and how it becomes leveraged by forecasting operations in the private and media sector.

To fully capture how weather forecasters make decisions, however, it is not enough to study the production of weather forecasts—one must study their consumption as well. This is true of all experts of course, even those experts working deep inside the hallowed halls of science. Scientific credibility, as Thomas Gieryn (1999) and others have shown, may be negotiated within a community of experts, but it hinges on how claims of expertise fare in the outside world. The considerations that go into articulating a knowledge claim, therefore, do not end with abiding by internal norms and conventions but must also take into account the expectations and interests of lay audiences and stakeholders. But whereas some groups of experts are only exceptionally required to strategically frame their expertise to mobilize lay support, other groups, of which weather forecasters are a prime example, have been assigned the "dirty work" (Hughes 1958, 122; Abbott 1981) of always formulating their decisions in the glare of public scrutiny. Charged with protecting the public from the vicissitudes of the weather, NWS forecasters take it as a matter of course that their pronouncements about the future will be eagerly seized on by great numbers of people, and they carefully fashion their predictions accordingly. To fully appreciate the considerations that go into meteorological decision making, then, it becomes paramount to also follow forecasters' claims downstream, beyond the weather forecasting operations room and into press rooms, workrooms, and living rooms (cf. Gieryn 1999). It was because of their central role in shaping the public image of weather forecasting that I turned to broadcast meteorologists about their forecasting process. Yet, to properly interrogate the analytic triangle of forecasters–weather–forecast users, I had to move outside the weather enterprise altogether and attend to the lived experiences of actual users of the NWS forecast. During my 2003–2004 visit at Neborough, therefore, I conducted a total of fifty-nine interviews with commercial fishers in two nearby fishing communities. Fishing is a highly weather-sensitive occupation, hence fishers are highly sophisticated users of weather information. And because it is only the NWS that issues marine forecasts, fishers directly rely on the NWS to decide whether or not to go fishing. Including their voice, therefore, allowed me to determine how NWS conceptualizations of their publics and their weather needs, as reflected in the day-to-day production of the NWS forecast, fare outside the walls of the Neborough office.

INTRODUCTION

Overview of the Book

This book advances a sociology of decision making by weaving together perspectives from pragmatist theory, STS, cultural sociology, organization studies, and cognitive psychology. Accordingly, it conceptualizes decision making as a fundamentally practical activity that relies on available heuristics, techniques, and resources—as determined by both the objective at hand and the evolving material and symbolic context of action—to fashion a provisionally coherent solution to routine and nonroutine challenges. Decision makers operate within a more or less institutionalized environment that, over time, affords them a stock of knowledge and repertoires for action, intended to help them respond efficiently to the microcontext of the decision-making task. However, I argue, it is within this evolving microcontext of action that decision making takes form first and foremost. Decision making in action is neither perfectly rational nor is it perfectly routinized. Instead, decision makers become skillfully improvisational and discerningly creative as a matter of habit.

To gain entry into the process of decision making, I have chosen to examine *expert* decision making in action not only because it has received the most explicit attention by scholars but primarily because it has received the most explicit attention by its practitioners who, in an effort to optimize their performance, have sought to externalize their process of diagnosis and prognosis into techniques and communicable rules of thumb, which are easily amenable to sociological study. More specifically, I have chosen to examine NWS forecasting in action—a highly calibrated, government-run decision-making enterprise about an iconically "chaotic" natural phenomenon with complex social ramifications. I thus follow NWS forecasters as they work the forecast routine, distill atmospheric complexity into a coherent diagnosis and prognosis, negotiate the accuracy and timeliness of their performance during ordinary and extraordinary weather events, anticipate the weather against different time horizons, and endeavor to relate authoritative and actionable pronouncements about the weather to their multiple audiences.

Chapter 1 situates NWS forecasting operations within the broader field of weather forecasting in order to illuminate the changing institutional logic underlying meteorological decision making at the NWS today. Centered on the controversy surrounding the recent NWS operational transition to a digital graphical forecast and grounded in the Neborough experience, it showcases how environmental forces can and cannot regulate decision-making practice.

Chapter 2 zooms in on forecasting life at the Neborough office. It takes the reader on a tour of Neborough's ecology, operations, and culture to ultimately settle into a discussion of the basic routine of a forecast shift—from the moment the incoming forecaster gets briefed by the outgoing forecaster to the moment she releases the NWS forecast to the world. While thoroughly intertwined in practice, the three main components of the forecasting task, to be considered in turn, are data analysis, deliberation, and forecast production. This step-by-step breakdown of the forecasting process sets the stage for the sustained examination of particular aspects of meteorological decision making in the chapters to come. But it also fleshes out and elaborates pragmatist theory of action with the day-to-day realities of diagnosis and prognosis at the NWS.

Chapter 3 takes a more systematic look at how NWS forecasters take stock of the weather and establishes that, in response to the need for situational awareness, they have cultivated an omnivorous appetite for information, routinely contending with a barrage of partial, crisscrossing, nonoverlapping, and often contradictory weather reports derived from a variety of sources. In fact, the weather in its tangible materiality still plays a consequential role in NWS forecasting practice, most notably when forecasters feel compelled to enlist their personal observations of the weather outside in the hopes of creating order out of the ambiguity and complexity of the weather on their screens. To illuminate how forecasters harness diverse information as they endeavor to project themselves into the future, I introduce the concept of *collage*, a heuristic that frames meteorological decision making as a process of assembling, appropriating, superimposing, juxtaposing, and blurring of information. Weather forecasting as the art of collage underscores the *culture of disciplined improvisation* that characterizes NWS forecasting operations. And it externalizes into *screenwork* and digital compositing, the cognitive labor of combining and distilling complex atmospheric data on hand into a provisionally coherent decision about the future of the atmosphere.

Chapter 4 amplifies the discussion by compounding the complexities of taming atmospheric indeterminacy with the particularly sensitive complexities of hazardous weather forecasting. Hazardous weather forecasting epitomizes the primary directive of the NWS to protect life and property. As such, it carries with it additional performance pressures. It is in the context of hazardous weather forecasting that NWS definitions of acceptable risk get worked out. The chapter begins by outlining NWS efforts to articulate and streamline the management of meteorological risk into a government meteorology. To properly showcase the challenges of actually managing meteorological risk on the ground, the chapter next launches into a thick description of two missed weather events during my fieldwork at the Neborough office. The first incident serves as an illustration of the serious repercussions that may arise from a missed hazardous weather forecast even in the absence of any hazardous weather whatsoever. The second incident details the aftermath of a forecast that missed the first snow squalls of the season, a forecast that would have gone unnoticed only a few days later but that on that day led to hundreds of road accidents and a statewide, five-hour-long traffic standstill. The concluding discussion, on the duality of meteorological error, reflects on the essential irresolvability of the "overforecasting versus underforecasting" dilemma and considers its implications for managing risk at the NWS.

Chapter 5 turns to the temporal dimensions of meteorological decision making. As a preliminary step toward temporally embedding weather forecasting practice, I identify two principles that underlie its logic: risk and spatial scale. The former rests on a demarcation between routine and nonroutine operations, while the latter is driven by the fact that the more global the reach of a weather phenomenon, the earlier its detection. The joint influence of risk and scale on weather forecasting practice yields four temporal regimes—and, I argue, four distinct styles—of decision making: emergency, extended alert, near term, and longer term. To flesh out and elaborate this rudimentary framework, I analyze its empirical manifestation in summer weather forecasting, winter weather forecasting, short-term forecasting, and long-term forecasting, respectively. In so doing, I complicate dual-process models of cognitive processing by establishing that, in practice, deliberation and heuristics are combined across disparate temporal regimes to produce organizationally sanctioned, skilled predictions.

Chapter 6 delves into the challenges and considerations that go into distilling complex and uncertain information into a relatable but still authoritative and actionable forecast message. It begins with an overview of the methods the NWS uses to assess the relevance of its predictions to the weather needs of the American public, paying particular attention to how conceptualizations of such needs are produced and reproduced in the interactions between Neborough forecasters and their multiple audiences. The "ivory tower" of the NWS, however, quickly gives way to the realities of commercial fishing. My interviews with Neborough fishers confound any easy distinction between meteorological experts and laypersons and expose the "deficit" model on which the NWS message is based. Paradoxically, the extent to which the NWS forecast is considered good by its users has little to do with its accuracy. Fishers' accounts of their weather-related decision making thus provide the final link between the production and the consumption of meteorological judgment.

Instead of a conclusion, the book's last chapter, chapter 7, explores the applicability of the overall argument beyond the case of NWS forecasting. It begins by schematically articulating how the main analytic components featured in the earlier, empirical chapters become entangled in the course of making decisions. Next, it theoretically extends this decision-making model along three analytically distinct dimensions: practice, temporality, and risk. Finally, it draws on the extant literature to offer some evidence for its external validity in two other, suitably different, decision-making fields—medicine and finance. Each one of these well-researched cases is meant to probe the analytic fruitfulness of the proposed conceptual framework along a different dimension. In the context of medicine, the focus thus shifts from prognostic to diagnostic decision making and from preventive guidance to intervention or remediation, while in the case of finance, the focus shifts from natural to social phenomena and the performativity of judgment and decision making.

ONE

The Weather Prediction Enterprise

Norm is in the process of publishing his forecast. The office bandwidth cannot handle sending out all thirty-nine forecast elements at once, so he has broken down the task into five chunks, "just to be safe." He is already half an hour past the dissemination deadline. Just then, a request comes in from the neighboring office to the north via the chat room: "Can you change the speed of your sky cover so we come into better agreement tomorrow?" Norm pulls up the neighboring forecast on his left screen, muttering to himself, "Join us at the thirteenth hour, why don't you." He turns to look again at model guidance and satellite imagery on his right screen: "Where the hell is he getting that?" With a long sigh, he types in the chat room, "Can I do it? Yes. Do I think it's going to happen? That's a different answer. I'll push the sky cover to 40% at [7:00 a.m.] and go from there." He goes back into his forecast to "fix" the sky cover. Turns to me: "That's some fine meteorology right there. I hope you're writing all this down." About twenty minutes later, over an hour past the dissemination deadline, he starts publishing his forecast all over again. Suddenly, out of nowhere, he bangs the table with both hands and exclaims at his computer screens, his face contorted with rage: "Come on, damn it, move!" He's been really agitated with the new forecasting technology all day, but I have never seen him be violent before. Everyone on the operations deck turned around startled, then quietly resumed their duties.

The following day, "the Norm incident" is making

the rounds in the office. Part of the fascination appears to be with how "thrown off" I had been by Norm's behavior. Apparently, I jumped up in my seat and then got very quiet for the rest of the day. People tell me these kinds of outbursts are not out of the ordinary: "you'll get used to them if you stick around." Peter is filling in Simon on what happened, adding in a grave voice, "everybody deals with their frustration differently."

Simon: [*Shaking his head*] It takes me, who knows the system inside and out, two hours to issue a warning. That's not right! People are going to implode at this desk one day. We need more bandwidth. It's ridiculous!

Peter: It's not just the bandwidth, though. Or all the bugs. Or the collaboration issues. Some of the stuff we do with [this technology] is almost beyond us. There is no skill. There is no way of verifying. That hurts our credibility. Who are we supposed to be forecasting for with this thing?

* * *

To begin to understand the process of decision making, one must begin by understanding the more or less institutionalized field of action in which it is fundamentally embedded. What is at issue here is not that decision making is social action but that it is social action precisely because it is *practical* action. Meteorological decision making would not be possible, let alone successful, without recourse to some notion of what counts as a weather risk, specific tools and techniques for predicting such a risk, or a sense of what constitutes a good weather prediction. And the same holds true for any kind of problem-solving or decision-making task. Yet, under normal conditions, the character of decision making as a social phenomenon has a way of receding into the background. Even within organizations, constituted precisely in order to tackle complex tasks most efficiently, it is easy to lose sight of the fact that every decision represents a communal achievement. This is because much of that communal work is typically unremarkable and invisible, hidden in the accepted standards, technologies, and rules of thumb we use to guide our judgments and decision making. In fact, the more invisible and taken for granted the environment in which decision makers operate, the more far-reaching its role in decision-making practice. But the more far-reaching its role, the more streamlined, efficient, and masterful decision-making action is bound to be. It is thus a testament to the importance of the decision-making environment that it is typically hidden from view, the center stage occupied instead by individual actors intrepidly confronting challenging and risky situations with great skill and resourcefulness.

It takes an extraordinary event to reveal the ordinary, invisible work that decision-making standards do for a community of experts. And indeed, it took a crisis for me to properly appreciate the hold their tools have on NWS forecasters. This crisis was not precipitated by an accident or some other decision-making failure—it was precipitated by a change in equipment. Soon after I arrived at the Neborough office, NWS forecast offices switched from producing a text weather forecast to producing a graphical weather forecast. Already contentious before its official launch, the new forecasting technology now became a daily source of struggle and frustration, as the above excerpt from my field notes illustrates. So much were forecasters accustomed to particular established ways of doing and reasoning that the implementation of the new technology produced a profound disorientation, extending to the very logic of decision-making action. A closer look at the institutionalization of the new NWS forecasting routine, therefore, offers a rare view into the normally hidden environmental forces regulating the process of meteorological decision making.

It is to that task that I devote this chapter after situating NWS forecasting operations within the field of U.S. weather forecasting. Through an examination of the unfolding and eventual closure of the recent NWS controversy, I introduce readers to the institutionalized environment in which NWS forecasters operate and to the professional norms, practices, and technologies through which its logic becomes articulated on the ground. By analytically grounding the discussion in the experiences and point of view of Neborough forecasters as they underwent this by all accounts painful operational transition, the aim is to provide a balanced perspective on how institutional factors do, and do not, influence decision-making action.

Enter the National Weather Service

Historical studies of meteorology go a long way toward furnishing the necessary framework for understanding the forces that gave rise to and continue to shape weather forecasting as a system of expertise. The weather has of course always struck fear and awe in the hearts of men: efforts to reign in its power by making it predictable have existed since before the beginning of recorded time, with the earliest known effort to systematize and theorize atmospheric physics being Aristotle's *Meteorologica*. But it was not until the late eighteenth century—and the spirit of Enlightenment sweeping Europe and the United States—that weather forecasting as an enterprise came into being (Golinski 2007). The catalyst was a relatively steady supply of weather reports as taking meteorological observations turned into a gentlemanly pastime (Jankovic 2000). Soon, local initiatives became more or less absorbed into scientific societies, such as the Royal Society of London or the Smithsonian Institution, and, aided by the adoption of the telegraph, there emerged a stable, albeit thin, network of weather observers spanning commercial telegraph stations, military hospitals, army posts, school academies, and colleges (Fiebrich 2009). It was the mounting demands for financial and human resources by these expanding weather observation networks around the globe that led to the establishment of dedicated, national weather agencies and the standardization of weather instruments and measurements in the mid-nineteenth century (Whitnah 1961; Hughes 1970; Fleming 1990). Data alone, however, did little to establish weather prediction as a science. If anything, the relative ease with which a wide variety of data on the weather could be collected intensified jurisdictional wars in the form of debates over the scientific merit of local weather versus global atmospheric systems, observation versus speculation, and reportage of unusual weather phenomena versus regular weather records (Jancovic 2000; Anderson 2005). Nor was the development of numerical weather prediction models in the early twentieth century enough to legitimize weather forecasting as a science—such efforts were in fact greeted as premature and misguided at the time (Friedman 1989; Lynch 2006). It took until the midcentury, when observing networks and numerical weather prediction modeling were harnessed to the computing power of machines—all thanks to the close links between the meteorological community and the military galvanized during the Second World War—for meteorology to emerge as a scientific profession in its own right (Harper 2003, 2008; Lynch 2006). And it was not until it had risen to the status of a science that meteorology was able to unify the communities of observers, theorists, and forecasters under one discipline—that of predicting the weather (Nebeker 1995).

This laconically brief historical outline of the professionalization of weather forecasting hardly constitutes a complete explanation of the phenomenon, of course. Equally importantly, it glosses over the development of the distinctive national traditions of weather forecasting that in their interaction brought forth the current "epistemic culture" (Knorr Cetina 1999) of the discipline. It does, however, help highlight key interrelated regularities structuring the field: (1) a long-standing and extensive weather observing network buttressed by a project of "infrastructural globalism" (Edwards 2006, 2010), (2) quantitative forecast models generated by computers powerful enough to process the massive amount of available data, and (3) wide-ranging and keen stakeholders given the universal

CHAPTER ONE

relevance and potential destructiveness of the weather. It is this last that has provided meteorologists with the collaborative data- and computer-intensive environment necessary for countering uncertainty, long before most other decision-making fields were able to muster similar resources and support. Nowhere are these three elements better aligned or more pronounced than in the operations of government weather organizations. Enter the NWS.

The NWS was established in 1870, just one year after the system of weather telegraphy—begun in the 1850s at the Smithsonian Institution but severely disrupted by the Civil War and a fire—was restarted once again at the Cincinnati Observatory (Fleming 1990, 141–62). Originally named "The Division of Telegrams and Reports for the Benefit of Commerce," it was nonetheless assigned to the Army Signal Corps and placed under the Department of War because "military discipline would probably secure the greatest promptness, regularity, and accuracy in the required observations" (Cox 2002, 95). Twenty years later it became a civilian organization, renamed Weather Bureau, and transferred to the Department of Agriculture, where it remained until 1940, to be transferred again to the Department of Commerce because of its importance for the growing aviation industry. In 1965, the Weather Bureau took on its current name and was assigned to the Environmental Science Services Association, which, five years later, became NOAA, the National Oceanic and Atmospheric Administration (White 2006). Today, the NWS is the world's largest meteorological organization, with about five thousand employees, a budget of close to a billion dollars (NWS 2012b), and an average of four hundred thousand weather bulletins a day.

Per its mission statement, the NWS "provides weather, hydrologic, and climate forecasts and warnings for the United States, its territories, adjacent waters and ocean areas, for the protection of life and property and the enhancement of the national economy. NWS data and products form a national information database and infrastructure which can be used by other governmental agencies, the private sector, the public, and the global community." To fulfill this mandate, the NWS maintains 122 forecasting offices around the country assigned to—and located in—a specific geopolitical region. It is these offices that are responsible for assembling weather data, creating forecasts and warnings for a prescribed area of forecasting responsibility, and disseminating weather information as appropriate. And it is these offices that allow the NWS agency to call itself "The People's Weather Service" and to justify its existence as "Your Weather Service" to governors and taxpayers alike. What is commonly understood as the NWS forecast, then, actually involves over a hundred separately issued and disseminated NWS forecasts, each catering to a particular area of forecasting responsibility but digitally stitched together into an apparently seamless national whole.

To be sure, forecasting at the NWS is as "big science" as it gets: the most modest and localized of pronouncements about the weather necessitates a massive coordination of people, resources, and technologies. Weather balloons are simultaneously launched twice daily from hundreds of locations around the world in order to sample the lower and upper atmosphere; hundreds of automated weather stations take minute-by-minute surface weather observations; hundreds of human volunteers continue to do the same at least once a day; over a dozen computer models are run multiple times daily, digesting the above and more observation data and transforming them into forecast guidance; hundreds of meteorologists at the NWS alone assimilate all this information to meet the varied weather needs of an entire nation.

The NWS solution to this massive coordination problem has been to surround its local forecasting operations with regional and national "Support Centers," with the majority of resources being concentrated under the auspices of the National Centers of Environmental Prediction, such as the Environmental Modeling Center, the Hydrometeorological Prediction Center, the Storm Prediction Center, the National Hurricane Center, and the Ocean Prediction Center. A typical NWS forecast may thus be produced by a single meteorologist stationed at one of 122 field offices, yet it embodies a tremendous organizational achievement. Within such a scheme, what counts as a good forecast will not infrequently be at odds with what forecasters themselves might consider appropriate, so that NWS forecasting practice effectively embodies a ceaseless negotiation among various competing logics of telling the weather.

Against this background, the decentralized organizational solution adopted by the NWS looms large and demands further attention. For it is indicative of a profound and deep-seated mind-set, pervading the entire agency, that sees it as unavoidable, if not necessary, that forecast offices be allowed to self-determine how operational directives should actually be operationalized in their jurisdiction. So much so, in fact, that, upon comparing the cultures of three forecast offices, Fine (2007, 71) arrives at the conclusion that in "the local offices of the NWS, it is almost as if 122 organizational experiments are running simultaneously." The rationale given by both NWS administrators and forecasters as the basis for this state of affairs is the same one presumably militating for the existence of field offices in the first place: the local particularities of weather and the particular weather requirements of local communities. Crucially, therefore, forecasters are deferred to not simply because they are presumed to be experts of the local indeterminacies of weather but also because they are presumed to be experts of its publics as well. Indeed, what gives them an edge over other government employees is not that they are scientists but that they are *public* scientists. That is why they ostensibly have the final word on the NWS forecast. That is why computer model forecasts but also forecasts from the Storm Prediction Center are expressly meant to serve as guidance only.

Not that NWS forecasters actually are public scientists. To be sure, NWS forecasting has a direct effect on the general public and as such is a "public-domain science" (Collins and Evans 2002, 2006). But it no longer satisfies the more restrictive definition of a "public science" because its primary audience is no longer the general public.[1] While NWS forecasts and related data are—still—freely accessible over the Internet by any member of the public, it is in fact the media and the private weather industry that serve as the primary producers of the weather information actually consumed by the American public today (Pew Research Center 2011). Certainly it falls on the NWS, qua government agency, to provide the official weather story of what actually transpired, and only the NWS may issue public alerts about impending hazardous weather. Yet long gone are the days when the media simply served as mediators of the NWS word. With weathercasters now expected to have a college degree in meteorology,[2] with a substantial amount of NWS meteorological guidance publicly available and most other model and observation data easily procurable at a nominal fee, it has become increasingly difficult to defend the claim that the weather forecasts featured on broadcast media are reproductions of the NWS forecast. Indeed, the television meteorologists I interviewed in the Neborough media market countered with bemused indignation any insinuation that they might be directly working off the NWS forecast, talking instead in great detail about their process of forecasting.

The shift from a quasi-monopolistic to a more competitive industry environment was already in progress in the late nineteenth century when private investors began realizing the economic value of weather forecasting services (Craft 1999). Private sector meteorology became properly established right after the Second World War, when, quite independently from each other, several former military meteorologists saw a business opportunity in offering to weather-sensitive industries the kind of very customized and specific forecasts they had been generating for weather-sensitive military operations (Spiegler 1996). But the private sector truly exploded with the "computer revolution" of the 1980s and the relatively cheap availability of meteorological instrumentation (Spiegler 1996, 432).

Today, no self-respecting television news station does not have—and appropriately advertise—its very own state-of-the-art weather radar. Having solved the problem of tools that effectively served as the gatekeeper of this "big science" occupational field, the private weather sector has grown exponentially in the last decades. Selling the weather pays, per the success of the Weather Channel and the persistent trends in public opinion surveys. Even as weather companies and media outlets compete among themselves for the attention of the American public, the NWS forecast has been increasingly forced out of popular print and broadcast media, with its last direct foothold lost in 2002, when the Weather Channel decided to start producing its own forecasts for its "Local Weather on the Eights." In 1995 the NWS divested its agricultural weather program and its (non-wildfire) fire weather program to commercial interests, the latest example of its policy obligation to give "due consideration" to the abilities of the private weather sector (NOAA 2006).

In light, then, of the general trend toward the propertization of public science (Nowotny 2005) and its particular manifestation in how the NWS may pursue its public science mandate, it becomes especially important for understanding the process of meteorological decision making at the NWS to follow the NWS forecast both upstream and downstream, during its verification but also during its consumption. Nonetheless, approaching the study of the NWS *as if* its forecasts are primarily consumed by the general public has much to recommend it. First, as a sensitizing concept, the notion of NWS forecasting as public science encourages a sociological interest in the changing landscape of doing science in the public domain and lays the groundwork for assessing and reformulating policy objectives. Although such a line of inquiry is beyond its scope, this book helps make the case that there are compelling reasons why the demarcation between private and public science, while constantly challenged, is tenaciously maintained. One can certainly not equate the two sides, however arbitrarily defined they may be. Second, the notion of NWS forecasting as public science serves as an important heuristic tool for appreciating the "politics of representation" (Mehan 1993; see also Goodwin 1994) through which NWS definitions of meteorological risk gain supremacy, indeed orthodoxy, over all others. As Fine (2007) shows in his discussion of how NWS forecasters become authors of—and hence authorities on—the storm, critical for the success of this project is the concurrence of doing science and doing government work. Unlike the fragile power of soft-money scientists who represent a reserve labor force for the government (Mukerji 1989), NWS forecasters' epistemic authority has been bolstered and secured by virtue of directly working for the government to protect the nation from the vicissitudes of the weather. Downstream, the deep uncertainty surrounding weather predictions and the heavy burden of impression management attached to any fortune-telling enterprise are considerably mitigated by large-scale infrastructure and the institutional veneer of a government seal. Within the walls of the NWS, competing interests and agendas come together under the overarching institutional logic of serving the public interest. The relatively flat, decentralized structure of decision making at the NWS thus both supports and is supported by the taken-for-granted assumption that the NWS is primarily, if not directly, consumed by the American public. To be sure, NWS forecasters must still cede their professional autonomy to the de facto superiority of bureaucratic desiderata, as is expected of government employees. Yet, in the name of doing public science, these government employees are afforded an unexpected amount of elbow room to define how the NWS mission is to be pursued on the ground, above and beyond the wide discretionary power that is typical of other "street-level bureaucrats" (Lipsky 1980). A third, and arguably the most compelling, reason for holding on to the notion of NWS forecasting as public science, then, is that this is also how the NWS represents itself to itself. However illusory the premise that NWS forecasts are primarily consumed by the general public, it nonetheless acts as the organizing principle—at once material and symbolic—of how NWS administrators and forecasters realize who they are and what they do. Such an organizing principle is not based on some sort of rational calculation of the expected utility of, say, securing public funds. Rather, it is driven by the "logic of appropriateness" (March and Olsen 2006) that the embeddedness in an ethos of government service inspires.

But you do not need a weatherman to know the winds are shifting. Unmistakably, weather prediction has become increasingly more market driven (Randalls 2010). And while cultural understandings of weather information as a public good (not to mention the difficulty of assessing the economic value of weather services) act as social limits to its commodification, one need only look at the privatization or self-financing regimes of the national weather services in Australia, Canada, Great Britain, or New Zealand to appreciate the gravity of the situation for the NWS. Thus far, proposals put before the Senate to divest the NWS from the sole authority of providing warning services or, more recently, to restrict public access to NWS information have been consistently unsuccessful. Yet such pressures, in the current climate of fiscal austerity, have left the NWS scrambling to remain organizationally viable. And with organizational problems being but the flip side of technical problems (e.g., Latour 1983), the NWS administration has, first and foremost, sought to address matters through an ostensibly technical solution: radically revamping its forecast production process. Not surprisingly, those most affected by the changes are, as ever, local forecasters, tasked with carrying out the public mission of the NWS on the ground.

I will have much to say in the next two chapters about how forecasters today actually produce the NWS forecast. In the meantime, for the insight it offers into the environmental forces structuring the process of meteorological decision making at the NWS, I turn to consider in some detail how the controversial new forecasting technology and, with it, the new logic of NWS forecasting ultimately prevailed.

Modernization and Its Discontents

When the federal belt tightening of the Reagan administration came to an end, the NWS embarked on a massive campaign of "Modernization and Associated Restructuring" to overhaul its antiquated observing and forecasting infrastructure.[3] It upgraded and expanded its radar network, simultaneously upgrading and expanding the network of its forecast offices. Where there were once 52 NWS forecast offices,[4] roughly equaling one office per state, now each forecast office equals (at least) one Dual-Polarization Doppler Radar, for a grand total of 122 forecast offices, the requirements of the NWS radar network determining the size and topology of the NWS forecasting network. Divvying up areas of forecasting responsibility among offices was, of course, still very much driven by political considerations. There are good and ample reasons why NWS forecasting is fundamentally based on political meteorology, as will become apparent in later chapters, efforts toward a "seamless forecast" notwithstanding. Yet it is also true that technical considerations—in this case, the prohibitive cost of establishing wideband communication links between the radar and the forecast office networks—have then and now played an equally decisive role in the configuration of NWS operations.

Similarly upgraded and expanded in the early nineties was the NWS automated weather observing network, which now boasts of nearly a thousand automated weather stations across the country, taking meteorological measurements every minute and transmitting them to forecast offices via the NWS supercomputer.[5] But the jewels of the NWS modernization were the new forecasting workstations. With its unparalleled capacity to store, access, and display enormous amounts and types of information, the Advanced Weather Interactive Processing System, or AWIPS (pronounced "ay-wips" in NWS-speak), has become the nerve center of NWS forecasting operations. So much so in fact that forecasters are wont to divide the history of the NWS into the pre- and post-AWIPS era. It is AWIPS that truly anchored the NWS forecasting routine onto the computer and established the image of the forecaster busy at work studying weather graphics on an array of monitors in front of him. By all accounts, AWIPS was a welcome relief from the earlier computer workstation, called AFOS (Automation of Field Operations and Services), which was deployed to replace Teletype and facsimile machines. Introduced at the end of the 1970s, AFOS already represented 1960s technology: all of its graphics were in black and white, its looping capabilities were crude, and text editing was extremely cumbersome. Consequently, the AWIPS technology was greeted as a long overdue operational upgrade and wholeheartedly embraced by forecasters.

While key forecasting technologies received a complete makeover during the NWS modernization, the forecasting process was left untouched. The advent of AWIPS, throwing the door wide open to the information highway, certainly had an effect on the quality of the NWS forecast, but it did not affect the production of the NWS forecast as such. The latter had essentially remained the same since the beginnings of the NWS in 1870: forecasters would analyze available observation and—in later decades—numerical model data with an eye toward *textually* formulating a weather prediction for the upcoming days in their area of forecasting responsibility. Indeed, while writing and communication media evolved over the years, the text of the NWS forecast remained reliably the same, no different in its studied concision and dullness than the weather diaries surviving to us from the Middle Ages (Fort 2006, 19).

It was the advent of another technology ten years later that radically transformed the NWS forecasting routine and propelled NWS operations into the digital age of weather forecasting. Introduced in the early 2000s as another component of AWIPS, a natural extension of the existing forecasting regimen, the Interactive Forecast Preparation System, or IFPS, in fact ushered in a break with the old ways of forecasting. NWS forecasters today begin the process of forecast production as they always have—by reviewing the available data. But now, instead of a text, they are expected to come up with a graphical formulation of their weather predictions (Ruth 2002). Specifically, they import their choice of model graphics into the Graphical Forecast Editor, the centerpiece of the IFPS software suite. Using an assortment of editing tools, they then diligently "massage" this baseline graphical forecast into their idea of what the forecast should look like. Producing the NWS forecast now consists of creating sets of graphical grids. Each grid set corresponds to a seven-day forecast for a single weather element—such as temperature, sky cover, or probability of precipitation—while each grid encompasses the entire geographical area of an office's forecasting responsibility at an hourly 2.5-square-kilometer resolution. Through clicks and drags of their computer mouse, then, forecasters must assign a color-coded forecast value to every grid point of each weather element out to seven days. To convey the NWS forecast for the next twelve hours, for example, traditionally expressed in a couple of spare, verbless sentences (e.g., "Mostly sunny with a thirty percent chance of showers. Highs in the low sixties."), forecasters are now tasked with producing a forecast for over sixty[6] weather elements. Once complete, the forecast grids from each field office are transmitted to the central NWS server, where they are mosaicked together into the so-called National Digital Forecast Database, available over the Internet in graphical format or as raw data (NWS 2014). Crucially, once grid editing is over, forecasting is also meant to be over: the legacy NWS text forecast is now automatically generated from the graphical forecast.

Without a doubt, the digitization of the NWS forecasting routine embodies an anxious attempt on the part of the NWS to remain viable and relevant as a government agency. Pressured to justify its existence—and its government funding—in an increasingly competitive weather market, the NWS sought to recast itself as a high-quality wholesaler of weather information. Per the then NWS director, the IFPS was to provide "a seamless stream" of weather data that the private sector could then value-add and repackage according to the needs of their customers. To render itself indispensable, then, the NWS endeavored to make its Digital Database an "obligatory passage point" (Callon 1986) for acquiring accurate, comprehensive, and up-to-date weather information. Yet, while first and foremost designed to relieve external pressures, the Database was of course commanding internal politics as well: in order for it to succeed, it required the mobilization of the entire assemblage of sociotechnical actors making up the NWS enterprise, not least of whom are NWS forecasters.

As is commonly the case with technological change in organizations, however, NWS officials did not properly attend to the embedding of the IFPS in forecasting operations. Not only did they largely take for granted the cooperation and loyalty of their employees, they also took for granted the technology itself, trusting it to be self-evidently superior so that hardly any enrollment work would be required. Training was minimal, limited to a superficial operational overview. In the words of a Neborough forecaster otherwise very much in favor of the IFPS,

[They] didn't provide us with any training. They said "Here you go, make it work." And it's fallen on maybe two people in each office to get the thing up and running. They never really, in my opinion, never sold the whole vision, and they should have. Because it's a, I hate to use the word, but it's a paradigm shift for everybody. And they never sold that, especially to some of the people who have been here for twenty, thirty years, who don't want to change. I mean you saw that here, too. Which is a shame, that shouldn't be. I mean I think the vision is being promoted a little more now, but that should have been done a long time ago, before all this happened.

This is a common and to a certain extent inescapable challenge in any technology implementation project. Every technology design carries with it a particular script for its use, inscribed in the very logic of its mechanics (Akrich 1992). In their presumptuous naïveté and wishful thinking, managers often come to expect that by enforcing a particular technology, they have in effect configured its use and its users (cf. Woolgar 1991). Despite their obdurate material presence, however, and no matter how thoroughly inscribed with the interests of their designers they may be, technologies remain interpretively flexible (Pinch and Bijker 1987). Just as users cannot unilaterally decide what a given machine is good for, so, too, are machines unable to predetermine how they are going to be coopted by a given user. Rather, the process of technology adoption and use is one of mutual shaping and reshaping. To use Andrew Pickering's (1995) evocative term, it entails a *tuning* of people and instruments, a recursive iteration of resistance and accommodation by the end of which the heterogeneous set of actors will have been reconfigured into a workable alignment. At the NWS, under pressure to come into better alignment with the private weather sector, the administration eagerly attended to the tuning of the National Digital Forecast Database to the concerns and interests of external stakeholders. The presumption that such an alignment was already in place as far as its own workforce was concerned, however, turned the implementation of the NWS gridded forecast into a protracted and torturous affair.

Absent a coherent operational vision, the implementation of the new forecasting routine essentially translated into forecasters' trying to figure out by themselves how to get this technology to work. The challenge, in this respect, was not how to make the IFPS produce a better forecast but rather how to get it to produce a good enough forecast according to the conventional standards of what a good NWS forecast looked like— namely, a text, not graphics. According to these standards, the IFPS was failing miserably. Not only was it dauntingly unfamiliar and unintuitive, but the technology itself was also clunky, slow, and erratic, making it a belabored and time-consuming battle to graphically edit the grids in order to get a decent text forecast. To make matters worse, the text forecast, spat out of the minimally developed text formatter, was so choppy and incoherent that it required extensive editing, or "QC-ing," to become passable. Meanwhile, the official rhetoric kept assuring that the IFPS "will not change" but only "streamline the forecast generation" by reducing "the mechanical and repetitive composition phase [and by] giving the forecaster relatively more time to apply his or her knowledge and expertise to forecasting" (Wakefield 1993).

It was only after they were faced with strenuous, unrelenting resistance and forced to acknowledge the extent of its toll on forecasting operations that NWS leadership now added the caution that embracing the new routine would require a leap of faith, a "paradigm shift." At the same time, accommodations were made to allow for further development of the IFPS software suite before the official release of the Digital Forecast Database. Announcing what would turn out to be the first of two postponements, the newly appointed NWS director warned, "The NWS needs to implement a consistent approach on how we integrate the Digital Database into operations" (NWS Focus 2004). At long last, it was clear that the new routine had to be sold to the staff as much as to Capitol Hill. Presentations making "a business case for change" to forecasters and stressing the importance of the Digital Forecast Database for the continued survival of the NWS grew increasingly common (e.g., Rezek 2002). And office management was widely advised that "while knobology training will be required for this transition, more important will be psychological training and clear communication of the goals of using the IFPS. It is critical to emphasize that this forecast tool is not designed specifically to produce text products, but to give forecasters the ability to paint a more detailed and meteorologically accurate picture of the atmosphere" (Maximuk 2003). Resistance on the part of forecasters was now cast as inability or unwillingness to shift paradigms and see the IFPS for the superior forecasting method it truly was or at least could be.

One of the most vivid and consequential formulations of this so-called paradigm shift came from Tom LeFebvre, one of the masterminds behind the NWS digitization project: "It's a bigger paradigm shift than switching from typewriters to computers. . . . It's more like taking a novelist who's used to expressing his ideas in words and making him an artist and saying 'I'm taking away your computer now, and I'm giving you some paint and brushes and I want you to draw your ideas'" (McGehan 2002). Such imagery sensitively captures the deeply embodied nature of expertise, and it points to important questions about what forecasters can and cannot do

CHAPTER ONE

with graphics as opposed to words. Yet, by keeping the focus on the creative struggles inherent in a change of expressive medium, it diverts attention from the more basic and fundamental identity struggles that the digitization of the NWS forecasting routine brought to the fore. In fact, there is a strong case to be made that employing graphics instead of words as their forecasting medium did not present a challenge for NWS forecasters. The learning curve for producing a graphical forecast was certainly very slow and painful, but that was mostly the result of the haphazard implementation of the IFPS. From the very first, the epistemic culture of weather forecasting has been deeply visual (Monmonier 1999). If they cannot see it, weather forecasters cannot think, never mind talk, weather. As I discuss in the next chapter, meteorological deliberations at the NWS always occur in the midst of weather visualizations, computerized or hand drawn. More importantly perhaps, the graphical *communication* of the weather forecast is hardly a novel concept. From the newspaper weather maps of the 1880s (Monmonier 1999, 157ff.; Calvert 1899) all the way to the Weather Channel phenomenon of the 1980s (Batten with Cruikshank 2002), the weather forecast has been told in progressively more sophisticated graphical formats. In a very real sense, graphics have long been the preferred expressive medium of weather forecasting.

To be sure, given the prevailing rhetoric around the operational transition, NWS forecasters were themselves also bound to frame their anxieties about the digitization of the forecast routine as a problem of expressive medium. Fine, whose fieldwork at the NWS occurred right before the implementation of the IFPS, poignantly illuminates the apprehension of forecasters qua authors (see especially Fine 2007, 151–70). Yet, even as they were bemoaning the transition to graphics, the real issue on NWS forecasters' minds of course was not the mode of expression but rather creative control over the forecast. For, unlike AWIPS and the other technological innovations of the early nineties, which were implemented to address existing operational concerns, the IFPS was intended to pave "a new way of doing business." Not surprisingly, therefore, it was fought tooth and nail by a great number of NWS forecasters.[7] Add to this the haphazard and uncoordinated embedding of the new forecasting routine into field office operations, and the official release of the Digital Database had to be postponed for one year, from the end of 2003 to December 2004. It was not until the NWS administration worked out that what was truly at stake was not authorship but authority over the forecast and took steps to negotiate and redress the situation that forecaster resistance abated and the matter became closed. In the meantime, in the process of getting mangled together into the current configuration of NWS forecasting practice, the emergent tuning of administration, forecasters, and IFPS spoke volumes about the logic behind the allocation of skill, jurisdiction, and risk during the production of the NWS forecast. When I returned to the field in 2008, the new forecasting routine was all but taken for granted, had literally become routine. Yet back in June of 2003, as I was given the "grand tour" on my very first day at the Neborough office, the first topic on everyone's lips, uttered in sardonic glee, was that I was about to witness a veritable paradigm shift in NWS forecasting operations. Two months later, forecast offices officially began using the IFPS to issue the forecast, and I was presented with the rare opportunity to monitor the experiences and anxieties of NWS forecasters as they struggled into the digital age of weather forecasting.

Controversies in science and the professions have long attracted the attention of social scientists. Despite the wide range of research perspectives and epistemological commitments involved, the underlying rationale is the same: studying knowledge production in a state of instability can unlock normally black-boxed or otherwise hidden dynamics and actors at play. Since, no matter the trigger, system instability is invariably accompanied by controversy, studying how particular controversies arise and are settled holds the potential to bring into relief the often invisible institutional, cultural, economic, legal, and political forces shaping and sustaining a given field of expertise. A closer look at the controversy surrounding the introduction of the IFPS, therefore, promises instructive insight into the field of NWS forecasting, insight that will prove essential for appreciating the environmental parameters structuring the practice of meteorological decision making. As I shall show, the technological implementation of the IFPS not only wreaked havoc with established, finely tuned weather forecasting techniques. More fundamentally, it represented a challenge to dominant standards of what constitutes a good NWS forecast and therefore what constitutes a good NWS forecaster.

The Call of the Weather Service

During my early days in the field, I was intrigued to discover that all Neborough forecasters but one became fascinated with the weather and "knew" they were going to pursue a career in weather forecasting at a very young age. "To me, it's a paid hobby coming to work," one of them mused out loud. "It's funny—sometimes I cannot believe this is what I do for a living." And it is not only NWS forecasters who articulate such identity stories. I encountered the same childhood fascination with the weather and the same narrative of weather forecasting as a calling[8] in my interviews with television meteorologists and in the newsletter of the National Weather Association, the primary professional association of operational weather forecasters in the United States. This is the norm, or at least it is promoted as such, within the field of weather forecasting at large. And it appears to be bolstered by two equally powerful, mutually reinforcing cultural scripts: being passionate about the science of meteorology and being passionate about serving the public. The former carves an occupational space for the true meteorologist, at a safe distance from the "weatherman" or the "weather enthusiast." The latter marks weather forecasters—or, more formally, "operational meteorologists"—apart from research meteorologists. It is in this light that NWS forecasters' otherwise perverse fascination with hazardous weather can be comprehended. Their excitement about the possibility of a blizzard or a hurricane hitting their area of forecasting responsibility is fueled by the heroic vision that they will ultimately save the day by "nailing" the weather system early on and by alerting the public to the dangers involved. To be a NWS forecaster is to maneuver between and draw on both these narratives. The emotional distress caused by a busted forecast becomes tempered by the reminder to oneself and to others that the atmosphere is inherently unstable, and it soon gives way to the excitement of researching what went wrong. Finally figuring out an answer to what the atmosphere wants to do is not very rewarding if it will not make a difference in people's lives. Within the setting of government work, *either* doing science *or* providing public service bears the stamp of a bureaucrat, a label NWS forecasters vehemently object to.[9]

Given the pervasiveness of this identity story forecasters tell to themselves and to others, one predictably encounters the same dual mission of serving the science and serving the public in their narratives of what constitutes "forecast goodness" (Murphy 1993). Accordingly, a forecast is good if it provides both scientific *and* valuable weather information. In other words, it is not enough that a given weather prediction meet certain professional standards of scientific quality[10]—it must also be deemed valuable by its users. Sure enough, NWS forecasters hold deeply problematized views on the subject.

Forecaster: An accurate forecast adequately describes what's going on both generically and psychologically. For instance, you can say, "partly sunny." And perhaps the numbers show that the overall sky cover was sixty percent, you know, during the time period of interest. However, if people are wearing sunglasses through all that time and lathering on the suntan oil, the behavior is more as though it's sunny

for most of the day. "Mostly sunny," you know, despite what the numbers and the verifications say, "mostly sunny" is still probably a better forecast.
Phaedra: But you still put "partly sunny" in the [forecast].
Forecaster: Yes, if the best information I've got says we're going to have sixty percent opaque, yeah, I'm going to aim towards "partly sunny." And, yes, perhaps we do get sixty percent, but maybe it's not all sixty percent opaque, it's just enough nonopaque cloud cover. So, in fact, to the average person it's not "partly sunny," it's "mostly sunny". . . . So we're facing both sides: not just the numbers, but also the effect it has on people. . . . The Weather Service has traditionally been, the first word I want to say is "stuck," but I'm not sure that's exactly the right word, but, they've been steady on using numbers, and that's the scientific way, the way we were all taught when we learned the science—use numbers to prove your experiment and so forth. So, it's reasonable, you know, the verification scheme with numbers is reasonable. But I would submit there are times when it's not always accurate.

In the above interview excerpt, in the process of discussing what constitutes an accurate forecast, this Neborough forecaster inevitably moves from forecast accuracy to forecast goodness to forecast verification and back again. While deeply interrelated in practice, these are analytically separate issues that I will strive to untangle in chapter 4. For now, this excerpt serves to underscore the inherent complexity and messiness facing any aspiration to forecast goodness. The dynamics between scientific quality and value can indeed make or break a forecast. Successfully taming the tension between these two desiderata is therefore the mark of a good forecaster. As already noted, this balancing act was traditionally achieved by way of the written word. Deceptively simple and unadorned, the formulaically brief text of the NWS forecast nonetheless exacted considerable care and attention from NWS forecasters, who would painstakingly weigh every word used to ensure their message was relatable and yet did not compromise the quality of their forecast. Neborough forecasters, to my initial amazement, would get into equally heated and protracted debates about the merit of using "brisk" over "breezy" in the forecast,[11] for example, as they would about the amount of snow or the timing of a storm. Up until only a few years ago, Mark Twain (1892, viii) was indeed right when he quipped that "Weather is a literary speciality." By textually articulating what they considered to be an intelligible, timely, sufficiently detailed, and reasonably accurate forecast, NWS forecasters had mastered a balancing act that, they felt, kept the dynamics between forecast quality and value in check.

The advent of the digital age, however, with its promise of endless possibilities for managing and communicating information, brought back the matter of what constitutes a good NWS forecast with a vengeance. Illustrative of the general animus of the day are the results of a short survey conducted at the Seattle NWS office in 1997 asking the staff for examples of a good forecast (Colman 1997). Respondents were deeply divided. Echoing conventional wisdom, one group maintained that good forecasts are concise, with as little detail as possible, such as "Scattered showers and sun breaks." For this group, "most people would experience benign weather so the forecast would be accurate short and sweet for them.... To include the additional material for [the few who would experience a thunderstorm] was not worth the risk of confusion." The other group of respondents, however, would rather have a detailed and specific forecast, such as "Rain ending and becoming partly sunny during the afternoon. Except mostly cloudy with occasional showers continuing north of Seattle through early afternoon." But they were quite skeptical of the ability of a text forecast to get all this information across.

Enter the National Digital Forecast Database. While never disputing that a good forecast must provide both scientific and valuable weather information, it nonetheless challenges long-standing notions of forecast goodness by redefining what constitutes a valuable forecast. As epitomized by the National Digital Forecast Database, valuable weather information is to be found now *first* in the graphical forecast and *then* the text forecast—words are an afterthought, literally a by-product of the graphical forecast. But by redefining forecast value in favor of digital graphics, the new forecasting routine also redefined the target audience of the NWS forecast, now recast as sophisticated users of weather information interested in detailed and configurable weather data rather than a weather story. Left without a weather story to tell, NWS forecasters lost their finely tuned balancing act. Not only did they now have to learn to be editors of graphics instead of authors of words; they also had to learn how to produce a good forecast at a five-square-kilometer level of accuracy. To aggravate matters, this forecast had to satisfy newly imposed consistency thresholds with neighboring offices. But naturally, weather forecasters are going to be uncomfortable conveying a lot of detail in what they consider to be "meteorologically complex" situations, where they feel there is not enough skill or information to make a prognostication that specific. And they are going to be reluctant to compromise what they consider to be an accurate—*their* accurate—forecast in favor of an interoffice consensus. In the eyes of many of its forecasters, by explicitly recasting itself as a wholesaler of weather information, the NWS was taking the back seat in catering to the weather needs of the nation. Sacrificing accuracy for the benefit of some sophisticated "NWS stakeholders" can seem like a high price to pay, especially since the new routine left little time to digest observation and model data and come up with a reasonably accurate forecast in the first place. And forecasters often, especially when aggravated by the complexities of the new technology, took this new operational philosophy to mean that "pretty pictures" are important now, not meteorology or public service.

To paint a fuller picture of the shifting norms and standards of what counts as good NWS forecasting practice, I will now turn to the mixed reactions the IFPS received during its implementation at the Neborough office. This is not to say, of course, that the IFPS had the same reception across forecast offices and across forecasters. On the contrary, it is safe to assume that there was significant variation in the adoption of the new forecasting routine, and indeed my interviews at NWS Eastern Region Headquarters confirm as much. Still, while not "representative" in a statistical sense, the wide spectrum of responses to the IFPS I encountered at the Neborough office arguably provides a fairly good sense of the spectrum of responses of the NWS forecasting community at large. The point at issue, at any rate, are the shared understandings of how weather forecasting is, and should be, done. As competent members of their profession, Neborough forecasters offered rich and authoritative insight into the matter.

The Implementation of the IFPS and Its Aftermath: A View from the Ground

From the outside, the operational transition at the Neborough office was generally regarded as exemplary. Certainly the Neborough office director had made it a point to push his office to the forefront of the new developments: "We try to be on the cusp of the IFPS curve, although it is driving our staff a little crazy at times. But we feel in the long run it will benefit our office." Within the office, however, the matter of how exactly the new NWS operational philosophy is to be implemented was far from settled. Although all forecasters, however reluctantly, had accepted the fact that the IFPS was here to stay, the proper negotiation of a good forecast given the new operational guidelines remained very much an open question.

As is to be expected, the forecasters who identified with the new operational vision had the least trouble conforming to the guidelines promoted by the NWS administration. Yet, while their close alignment with the official position was sometimes derided by other forecasters as a sign of opportunism and a lack of commitment to professional standards, they were not simply regurgitating the official rhetoric but displayed a deep, reflective understanding of what was at stake.

Right now we're, the Weather Service is, having some trouble getting the forecast ownership—and to some extent even still today your name is at the end of the text, so it's *your* forecast—shifted from the words to the information that's contained within the words. And that, to me, that's what we're really here to do. It is to not necessarily convey the words "partly cloudy," but to convey the information that the "partly cloudy" conveys. Because, I mean, a lot of weather is subjective, you could ask ten people on the street, "Is it mostly sunny or partly cloudy today?," and you're probably going to get quite a few different answers. So the information to me has always been more important than the actual words. . . . The beauty of the Digital Database is, and the reason why I'm all for it is, "Here's the information the best we can provide, make your own decision." You know, even what we put in the text, it's, we just arbitrarily decided, okay, if it's not a thirty percent chance, we don't mention it. But that to me is what's wrong with just relying on text products to convey the information, because it is to me an arbitrary set of rules and, you know, that concrete contractor who's looking at this forecast, that may not suit his needs. But I know the information, I can get the information across visually and let them figure out for themselves whether they want to, you know, take the risk for themselves or not. I don't want to, you know, I can't be responsible for, "Well, there's a thirty percent chance of rain today, you really shouldn't pour your concrete costing ten thousand dollars" and then everywhere it rained except where they were. So now, you know, I can't be thinking about that as I'm going about my job: "Can't put a chance of rain in there because I'll cost people money" or. . . . That's not really what I think our purpose is here. It's to just "Here's the information, make of it what you will." And the [text forecasts] are so limited as to what you can convey in a concise fashion.

Hailing from one of the first offices to experiment with a gridded forecast, this forecaster had never issued a forecast without the aid of the new technology, something he considered an advantage over other forecasters "because I don't know any different." He devoted hardly any time to editing the generated text forecast compared with most. And he never signed his forecast, an omission unthinkable for the rest of the staff. Indeed, it would be fair to say that for this forecaster, and for others like him, providing the best public service in the digital age effectively means not having to worry about public service at all. As far as he was concerned, any negotiation between quality and value on the part of forecasters is bound to be based on "arbitrary," unscientific criteria. What makes the IFPS a superior forecasting system is that, by translating the gridded forecast into a wide variety of outputs, it does away with this compromised arrangement. The responsibility of interpreting the forecast is now displaced onto forecast users, while forecasters are released from the burden of making executive decisions on their behalf and, thus unencumbered, can concentrate on coming up with the most scientific forecast possible.

Such a position, however, was hardly going to be palatable to a professional group whose identity has long been bred and sustained by conceptualizations of a "Joe Public" not interested in deciphering a forecast but just wanting to be told whether it is going to rain tomorrow or not. Underscoring the deep cultural dissonance engendered by the NWS shift from a "push" to a "pull" model of forecasting operations, whereas the sizeable but relatively small number of IFPS advocates hailed from the post-AWIPS era, expressions of resistance and alienation toward the NWS operational transition cut across generational lines. I encountered several NWS forecasters in their midtwenties who, although more circumspect in their pronouncements, nonetheless shared the same strong concerns and negative sentiments about the operational transition as their more senior colleagues. To make matters worse, the technological growing pains of the transition kept everyone frustrated and overwhelmed, resentful of the toll the new routine was taking on their performance:

It's more of a resignation because it's going to happen, it's happening, we have no choice on the matter. . . . But the thing is that there's still a lot of resistance because of the fact that they just can't, you know, can't keep everything copasetic for at least a week and a half. [*laughter*] I mean I can understand change is needed, but it just seems like there's just too much all the time, and as soon as we get one thing absorbed, it just switches up and does something else.

While overt acts of defiance were quite rare and generally frowned on, "everyday forms of resistance" (Scott 1989) were a common and pervasive occurrence, typically articulated via the pursuit of what was framed as a forecaster's first priority: producing a forecast to be proud of. By the summer of 2004, over a year past the launching of the IFPS routine, producing a forecast to be proud of still meant that most Neborough forecasters would routinely work overtime, meticulously editing the text forecast for a full thirty minutes to an hour after having spent hours toiling over the gridded forecast. It also meant that, when only relatively minor updates to the forecast were necessary, forecasters would update the text but not the grids. And they might neglect a grid detail or even ignore the dissemination deadline so as not to compromise the quality of their forecast. The following interview excerpt captures the prevailing mood:

I want to be comfortable with my product out there because it is my forecast with my name on it, and I want to be comfortable with the forecast that I put out, whether it be written or gridded, and I want to feel comfortable with that forecast, okay? And that, I think, that's important. So, yes, 4:00 p.m. rolls around, and I know that I've got to get that down to the Digital Database. So I just send whatever I've got ready to go down there. Usually I've got most of it done anyway, but there's maybe a couple little tidbits I don't have done. But I'm to the point where I do the best that I can, and whatever I have done it goes, and that's it!

It should come as no surprise, then, that the directive from Headquarters to "Concentrate on the grids and forget the words!" was the object of much ridicule and derision at the Neborough office. Yet, whereas most forecasters clung to the text forecast in order to cope with the growing pains of the new forecasting routine, a small number engaged in what organizational behavior theorists Danny LaNuez and John Jermier (1994) call "sabotage by circumvention": subtle but highly effective forms of noncooperation intended to subvert institutionally sanctioned procedures and policies (see also Morrill, Zald, and Rao 2003). Rather than simply react to the daily challenges of the new forecasting routine, these Neborough forecasters were quite strategic in their appropriation of the IFPS, and they soon found ways to interfere with the new technology so that it performed according to the old standards of forecast goodness. One such stratagem, known across the NWS as "carpet bombing," entailed applying the same value over a large spatiotemporal area of a given weather element grid in order to control the text output. Carpet bombing effectively turns the IFPS on its head. One is no longer painting a map but a text on the computer screen. Uniformly coloring, for example, a large enough area of the Sky Cover grid for several time periods so as to obtain a text of "overcast in the morning" and then not including any cloud cover for a number of time periods so as to obtain "clear in the evening" does not reflect reality: clouds do not suddenly disappear, they gradually clear. One of the main arguments in favor of the gridded forecast is precisely its ability to convey atmospheric processes in more detail as it is composed of hourly high-resolution grid points. Not trusting (the politics of) the algorithms embedded in the new forecasting technology, these forecasters resorted to carpet bombing to preemptively manipulate the text output and ensure that graphics were blunted—all in the name of providing a good forecast, of course.

I can live with this less carpet-bombing policy once we get ensemble models, but we're given categorical model data to populate the grids, and that creates problems.

* * *

Gradations and probabilities of precipitation are hurting our forecast. Rain is a yes or no. . . . The customer won't understand what's going on. This preoccupation with carpet bombing is contemporaneous with the reality of the world. Temperature is continuous, precipitation is not. Drivers need to know if it's going to rain or not.

It is the same imperative of forecast goodness, with its slippery twin allegiance to scientific quality and public service, that triggered another common subversive forecasting practice. Traditionally, NWS forecasters would make no mention of a precipitation event if its predicted likelihood of occurrence was less than 20 percent. Since the introduction of the Digital Database and the push for greater detail in the forecast, however, the minimum probability threshold for including a precipitation event has been lowered to 15 percent. I have encountered several NWS forecasters who, even though they may estimate the likelihood of a given precipitation event to be around 15 percent, will nonetheless artificially keep the Probability of Precipitation grid to no more than 14 percent. This way, they are not required to paint a corresponding precipitation event in the Weather grid,[12] and the IFPS formatter will not generate a mention of precipitation in the text forecast. The motivation behind this practice is to not unduly alarm the general public who, these forecasters believe, will hear "slight chance of showers" but only retain "showers." Unlike most other forecasting tactics that surfaced as a reaction to the implementation of the IFPS but have since by and large disappeared, this one had survived with a vengeance when I returned to the field in 2008. Similar to the heated debates over using "blustery" versus "brisk," or "fair" versus "clear" (Fine 2007, 158–61), in their forecasts, NWS forecasters were now consumed in protracted discussions over the significance of drawing a 14 versus a 15 percent Probability of Precipitation grid point.

It may be tempting to attribute the persistence of this last practice to some sort of inherent merit that sets it apart from the politically motivated rest. And indeed, whereas most Neborough forecasters would become exasperated with colleagues' use of carpet bombing, they were sympathetic to any challenges to the new lowered probability threshold. As a rule, carpet-bombed forecasts were promptly amended by the next shift; artificially precipitation-free forecasts were not. Yet the evidence that alerting to the potential for precipitation at no lower than 20 percent—as opposed to 15 or 30 percent for that matter—of probability makes for a better forecast is far from conclusive.[13] More to the point, forecasters are not in the habit of basing their decisions on the latest such research. As I discuss in later chapters, commitment to public service, at the NWS and throughout the weather forecasting enterprise, has thus far not been accompanied by an equal commitment to the systematic study of the social effects of the weather and of weather communication. To explain the persistence, hence the legitimacy, of this form of sabotage by circumvention, one must instead look to the culture of NWS forecasting that imbues it with objective validity. It is thus reasonable to assume that this practice has stuck because it resonates with what still seems to be one of the cornerstones of the collective identity of NWS forecasters—namely, that they are there to serve a meteorologically unsophisticated Joe Public. Coupled with the ever-present challenge of quantifying multiple types of uncertainty under precipitation scenarios of varying difficulty, such an interpretation of their professional role and responsibility ostensibly gives forecasters license to sidestep a policy that threatens accepted standards of service. In the same vein, one must resist the temptation to attribute to some inherent flaw carpet bombing's failure to acquire staying power. The fact that Neborough forecasters dismiss carpet bombing as "bad science" is the *result* of its failure to find fertile ground in the master narratives and standards of NWS forecasting culture, not the cause. Unable to marshal widespread support, it could not withstand for long the "trials of strength" (Latour 1987) against the official forecasting routine.

Standards of professional practice, however, serve to redraw as well as police the boundary around truth and legitimacy (Bowker and Star 1999; Ottinger 2010). They are driven by a need to streamline and thus regulate knowledge production, but their robustness and longevity depend on allowing for actors with different commitments to work together successfully. Ideally, therefore, they must evolve into "organic infrastructures" (Star 2010, 602) such that the sociotechnical arrangements and normative positions originally instantiated by them will recede into the background, giving way to the situated logic of action and interaction. It thus takes a crisis or some other such form of "infrastructural inversion" (Bowker 1994) to reveal the otherwise invisible work that professional standards do for a community of experts. And since new standards prevail not because they are more rational but because they are backed up by a stronger network of allies, it takes a crisis to reveal the full cast of characters—humans, technologies, nature—with a stake in the developments. The introduction of the IFPS as the new standard for producing the weather forecast constitutes such a crisis for the NWS. As already discussed, the "boundary work" (Gieryn 1999) summoned to the fore by the ensuing controversy exposed a commitment to science and a commitment to public service as the basic building blocks at the core of NWS forecasting identity and logic of practice even as it fueled competing visions of how best to promote them. It is around this dual commitment to science and public service that the cultural map of NWS forecasting continues to be redrawn: when the IFPS was first introduced and, years later again, with the credibility wars over and the dust all but settled. To be sure, the boundary line demarcating sound from unsound forecasting has shifted considerably since the cultural map of the pre-IFPS era. The orienting landmarks now pay tribute to precision and teamwork rather than accuracy and individual performance. Still, what accounts for the normalization of the IFPS forecasting routine is not so much that it drove out carpet bombing, for example, as that it can accommodate a wide variety of forecasting philosophies and styles. After all, carpet bombing and other such stealthy forms of dissidence, although contributive, were hardly the main culprits in derailing the timeline for the official unveiling of the new NWS forecast. At the root of the matter, rather, was a profound discontentment and recalcitrance running rampant across forecast offices. A weakly defined, albeit organic, infrastructure works to keep the grumbles at bay.

For, needless to say, the NWS crisis was not only ripe with "ethnographic moments" to this sociologist's delight. It proved highly insightful for the NWS administration as well, who, in the process of trying—and failing—to institute a total system of weather forecasting, had to learn how to prevail over resistance and dissidence. Central in this effort was the IFPS technology itself, which, linked to the NWS supercomputer, could readily serve as an "information panopticon" (Zuboff 1988). Ostensibly, it was set up to track forecasters' adherence to the new interoffice collaboration guidelines, every day assigning a yellow smiley face or a green sickly face to each office on an internally public list. There can be little doubt, however, that the enormous volume of data churned out by the IFPS allowed administrators to pursue a rigorous bird's-eye view on all aspects of forecasting operations.[14] Such information was initially put in the service of drawing a hard line between right and wrong forecasting behavior, as is typical during a time of crisis. Yet this strategy was doomed to failure in a rather horizontally structured organization aiming for a fractal business model and with a long-standing tradition of resorting to "soft power" (Nye 2008) rather than harsh measures to secure its interests. The repeated postponements of the target opening day for the Digital Forecast Database along with an unsuccessful attempt to centralize forecast offices led, by all accounts, to the replacement of the then NWS director. Since then, the gray zone around what constitutes a good NWS forecast has gradually expanded again. The salient features of the NWS forecast still revolve around the Digital Database, and forecasters do not have final editorial control of the text that pops up on the NWS website, were one to request the forecast for a particular town. But the legacy county, or "zone," text forecast, available for forecasters to directly edit as they see fit, is always prominently displayed alongside the IFPS-derived text forecast, and it is this text that is broadcast by the NWS radio and other affiliated media. The ambiguity and contradictions of this operational borderland function as a safety valve, helping to preserve the agency's authority by providing a space for creative control over the forecasting process. Forecasters are able to flout official rules but still stay within the bounds of what is permissible, while the NWS is still able to profit from the skills and expertise of employees who repeatedly transgress the strictly defined boundaries of some of its policies.

When I returned to the Neborough office in 2008, any remaining rumblings against the IFPS seemed nothing more and nothing less than an interaction ritual, part of the "rhetoric of complaint" (Weeks 2004) of an occupational culture whose credibility is inextricably tied to the competence and incompetence of machines. At last, forecasting life had settled into a hectic but steady pace. Most Neborough forecasters now left the text forecast as is during benign weather—supposedly because the IFPS text-formatting software had "improved a hundred percent," as they told me, but more fundamentally because they had come to terms with the new status quo, as was evident to me. To be sure, their overwhelming insistence to see consumers of the NWS text forecast as their primary audience remained at odds with the official position on the matter. Yet, with usage data showing time and time again that the bulk of traffic to the Digital Database comes from individuals and not government and corporate entities, forecasters' conviction that they are there to serve the weather needs of Jane and Joe Public has been progressively reconciled with the official redefinition of the NWS forecast as a graphical forecast. Just as importantly, they have come to see the transition to the Digital Database as an inevitable necessity, echoing the neoliberal logic NWS administrators used back in 2004 to make a case for the IFPS: "We had to really change with the times and the market because, if we didn't, then we'd be going extinct, and private weather companies would take the whole shebang over, which they're still trying to do anyway."

In light of the kinds of external challenges facing the agency, it is quite possible that the primary identity marker of the NWS forecaster will shift squarely from protector of the general public to advisor to emergency managers within the next generation of forecasters. For the moment, however, these multiple identity orientations—and the forecasting practices they inspire—exist side-by-side on the culturescape of NWS forecasting. What was fanned into identity conflict and factionalism during the early stages of the operational transition and the imprinting of a new institutional logic on NWS forecasting practice has now settled into a more or less stable, nested identity structure that provides a rich set of resources for strategic action at the individual, group, and organizational level (cf. Owens, Robinson, and Smith-Lovin 2010). It is against this delicately rebalanced backdrop that meteorological decision making takes place at the NWS today.

Coda: The Subjectivity of Objective Weather Forecasting

There is no doubt that the NWS owes its status, credibility, and funding to the claim that its predictions are based on science—or, more precisely, that its predictions are based on so-called objective methods of determining the weather. It was not so long ago, however, that weather forecasting had no claim to objectivity. Well into the interbellum years, in fact, meteorology was disparaged as a "guessing science" (in Harper 2008, 52). The British Meteorological Office stopped issuing daily forecasts for over a decade in the late nineteenth century (Nebeker 1995, 39), while the U.S. Weather Bureau would not issue long-range forecasts until almost the middle of the twentieth (Harper 2008, 26; Pietruska 2011). Weather prediction was hardly beholden to explicit theoretical principles and a rigorous process of decision making, skirting dangerously close to being indistinguishable from the religious weather prophesying and other forms of meteorological divination of the time. Per a 1918 National Research Council report, "Practically all of the rules known and used in the art [of forecasting] have been established empirically; some of them have been formulated, but in a considerable proportion of cases the rules which govern the forecasters are exercised subconsciously" (in Nebeker 1995, 40).

It was the Second World War and the frustrations of trying to achieve consensus among fleet forecasters that fueled the search for algorithmic solutions to predicting the weather. "Objective" methods of data collection and interpretation were for the first time consciously juxtaposed to "subjective" methods, and the project of numerically predicting the weather by employing the brute force of computers to mathematically simulate the laws of atmospheric physics began in earnest (Nebeker 1995, 127ff.). Already by the midfifties, Lewis Fry Richardson's 1922 dream that "it will be possible to advance computations faster than the weather

CHAPTER ONE

advances" (in Lynch 2006, 1) was on its way to becoming a reality. The formulation of chaos theory by Edward Lorenz a few years later, with its profound implications for the predictability of weather, sealed the fate of numerical weather prediction and paved the way for simulating atmospheric dynamics. The modern science of weather forecasting had been established.

Today, the weather forecasting enterprise, at the NWS and elsewhere, is manifestly based on output from the "objective analysis" and forward rendering of the atmosphere by a variety of numerical prediction models. Yet claims to scientific authority and hence credibility bear a complicated relationship to claims to objectivity. On the one hand, there is ample evidence of the kind of scientism that has become increasingly hegemonic in contemporary mainstream culture and discourse. The official NWS rhetoric, whether for public or internal consumption, is replete with reductionist language and technocratic buzzwords, while forecasters readily subscribe to a naively positivist vision of science—even when, or precisely because, they keep an ironic distance from it (cf. Fine 2007, 77ff.). On the other hand, NWS operational guidelines explicitly and repeatedly leave it to forecasters' judgment and discretion how numerical prediction models may assist them in their task. Indeed, it is common practice in English-speaking meteorology circles to lump together all forecasting tools, including forecast models, under the term *guidance*, which may then become qualified into *model guidance, objective guidance, observational guidance*, and so on.

At a basic level, what allows for these opposing impulses to coexist, often side by side, in the world of NWS forecasting can be attributed to changing notions of scientific objectivity. As historians of science Lorraine Daston and Peter Galison (2007, 371) have persuasively argued, "there can be, there has been, there is science without mechanical objectivity." The pursuit of objectivity first became coupled with the pursuit of science in the mid-nineteenth century, the age of mechanical reproduction. Although pure mechanical objectivity was recognized as a chimera before long, the pursuit of objectivity continued to prevail as a defining feature of scientific activity, triggering a pluralization of the concept. Far from a universal constant, the ideal of objectivity—as a form of knowledge and as an epistemic virtue—has become exceedingly fluid, amenable to appropriation by distinct groups of actors. Daston and Galison (2007, 321ff.) examine in detail two such reincarnations of the ideal, both of which followed in direct response to the internal inconsistencies of the notion of mechanical objectivity, and both of which are relevant for the NWS case. The first response was arguably a more extreme version of the original. It sought to capture the objective—that is, the fixed and universal—laws of nature by substituting the rigor of mathematical and logical structures for the idiosyncratic and ephemeral, albeit machinic, renditions of it. The second response was to transform the erstwhile liability of subjectivity into an asset, in effect making it objective. Objectivity is recast as expert judgment, and mechanical reproductions of nature now require interpretation by a professionally trained eye to reveal what is meaningfully real. Reminiscent of debates currently preoccupying the NWS, objectivity is put in the service of accuracy, which is increasingly seen as socially contingent and hence in need of interpretation.

One encounters all of the above orientations to scientific objectivity at the NWS today: the pursuit of mechanical objectivity has militated in favor of replacing trained human personnel with automated observing stations; the proliferation of numerical prediction models can be attributed, in part, to an effort to compensate for the weaknesses of automated observing systems; and trained human personnel have been locally embedded to push past the interpretive limitations of prediction models. These conceptualizations of objectivity, among others, often serve as convenient fault lines around which actors rally and define themselves, but they keep reproducing themselves fractally throughout the field of NWS operations as one shifts one's gaze to within-group formations (cf. Abbott 2001a). Objectivity is employed as a gambit and invoked strategically to stave off threats to professional status—its meaning, therefore, is highly context dependent as well.

More often than not, of course, the preferred strategy is to avoid direct invocations of objectivity altogether. It is telling that with the exception of the almost "obsessive" referencing to objectivity and automation during the planning and publicity of its multibillion dollar modernization program in the nineties (Doswell 1990, 30), the NWS has in recent memory consistently eschewed direct claims to it, opting instead to advertise the accuracy of its forecasting operations (see chap. 4). Equally striking is the absence in the NWS operational literature of direct references to "objective guidance" or to any variation of the otherwise very common terminological distinction between objective and subjective or judgmental forecasting techniques. To all outward appearances, the NWS has rejected this dichotomy as a false one and resolved to transcend it by making it irrelevant. Indeed, by promoting a forecasting culture of disciplined improvisation, as I discuss in chapter 3, it has unmistakably endeavored to make a virtue of the fact that, to paraphrase the earlier quote by Daston and Galison, there can be, there has been, there is weather prediction without mechanical objectivity.

And yet the lure of objectivity as the surer, preferred path to forecast quality is lurking everywhere, the IFPS case being no exception. As we have already seen, in an effort to remain competitive in a highly crowded marketplace, the NWS has sought to redefine the field and reassert a central position within it by introducing the IFPS technology as the gold standard of weather forecasting. And while it still remains an open question how the boundary work and other "distinction practices" (Burri 2008) of the NWS will fare across the profession, there can be no doubt that, within the NWS itself, the institutionalization of the IFPS has changed the logic of operations in a subtle but consequential way. This is because the IFPS shifts symbolic and material emphasis away from expert judgment and toward mechanical objectivity. The redefinition of the NWS forecast as a graphical forecast effectively privileges model guidance, which now formally becomes the baseline of every forecast issued by the NWS. This in itself is not a novel approach to NWS forecasting—forecasters have always based their decision making on model guidance. By virtue of explicitly scripting into orthodox forecasting technique the use of model guidance, however, the NWS for the first time explicitly recognizes expert judgment as a liability, albeit an unavoidable one. Yet this recognition hardly signifies a return to the ideal of mechanical objectivity. Rather, as Daston and Galison (2007, 414) suggest, it points to the emergence of a new hybrid form of expertise that "combines [the] ethos of [a] late twentieth-century scientist with [the] device orientation of [an] industrial engineer and the authorial ambition of [an] artist." It is on this basis, as we shall see, that NWS forecasters are called on to master the uncertainty of the atmosphere.

TWO

Working the Weather: A Shift in the Life of a Weather Forecaster

The fact that, to truly behold meteorological decision making in action at the NWS, one must come at it from inside one of a total of 122 field offices carries with it nontrivial methodological challenges. Such challenges are, to a large extent, attendant to any ethnographic project, of course, but they are further compounded in this case by the decentralized organizational structure of the NWS. One of the main conclusions in Fine's (2007) study of the culture of NWS forecasting was precisely that it is highly localized, with each field office practicing meteorology in accordance with its own unique standards and lore—in accordance, in other words, with a distinct, albeit complementary, "idioculture" (Fine 2006, 1979). The implementation of the IFPS (Interactive Forecast Preparation System) since then has greatly helped attenuate local differences and bolster collectively shared ways of doing and thinking about the NWS forecast. As I will show in the following pages, the production of the forecast is increasingly shaped by interactions across, not just within, NWS forecast offices. The emergence of an interaction order spanning multiple neighboring offices, however, in no way diminishes the influence of the local. To properly appreciate meteorological decision making in action, one must first develop an appreciation for the environment—climatological, cultural, institutional, technological—in which it is *emplaced* (cf. Gieryn 2000). I therefore begin the introduction to the NWS forecasting routine with an introduction to forecasting life at the Neborough office.

Yet, even as I will be endeavoring to cast into relief what is distinctive about Neborough, juxtaposing it to the three offices in the Midwest studied by Fine, what is bound to also come to the fore is exactly how much NWS field offices have in common. After all, they are all expected to practice forecasting the NWS way. They are all, therefore, equipped with the same set of big and small forecasting technologies, technologies that, as we saw in the previous chapter, are designed to elicit specific, organizationally endorsed decision-making habits from their users. No less importantly, all NWS field offices have access to the same sources of weather information, they all are tied to the same reflexively structured coordination network (cf. Knorr Cetina 2003), and of course they all have been inculcated with the same institutional logic of weather forecasting service. As a result, the various office idiocultures are all part of a more or less cohesive epistemic culture (cf. Knorr Cetina 1999). Notwithstanding local differences, forecasting practice at the NWS is aligned around a common organizational process and shared core principles of meteorological decision making.

And that is why, for the purposes of this analysis, it is ultimately of no consequence that I chose to pursue an ethnography of NWS forecasting operations at the Neborough office as opposed to another office, although this was certainly an issue over which I agonized long and hard during the design phase of the study. If my distillation of the process of meteorological decision making is to be credible, it does not—and should not—matter in which NWS office I conducted my fieldwork. It is also true, however, that my selection of the Neborough office guaranteed a most exciting and rich ethnographic experience. If there was ever any lingering doubt that workplaces acquire their own distinct personality, the colorfulness of the Neborough office quickly puts it to rest. Neborough combines breathtakingly changeable weather year round, a long-standing vibrant meteorological community, and one of the biggest metropolitan centers in the country. What is more, the numerous fishing towns along the Neborough coastline and their direct dependence on the NWS for marine weather information held the promise of allowing me to study the production of the NWS forecast within the context of its consumption. And so it was with much nervous anticipation that on a Monday morning in late June of 2003 I found myself staring up at the Neborough office building that was to be my home away from home for the foreseeable future.

This chapter takes the reader on a tour of life at the Neborough office, zooming in on its forecasting structure and culture to ultimately settle into a step-by-step discussion of a typical shift. This breaking down of the forecasting process—from the moment the incoming forecaster gets briefed by the outgoing forecaster to the moment she releases her forecast to the world—is meant to provide a guide to the "grammar," if you will, of NWS forecasting practice and serve as a point of departure for the more elaborate analysis about specific aspects of the process of meteorological decision making in the chapters to come. By the same token, this chapter serves to bolster and flesh out the argument, that a pragmatist theory of action best captures the day-to-day realities of diagnosis and prognosis at the NWS.

Location, Location, Location

The Neborough office provides warning and forecast products and services for about thirty-five counties in the northeastern United States, covering a population of roughly eight million residents. Its establishment in 1870 makes it one of the oldest forecast offices of the NWS. As such, it was until recently located where the majority of its publics work and play: Metrocity, the urban capital of its area of forecasting responsibility. The expansion of the NWS radar network in the early nineties, however, dictated the relocation of the Neborough radar—hence the move of the Neborough office to its current location in an industrial park about fifty miles away from Metrocity.

The selection of a suitable relocation site was not an easy matter. Besides the obvious issues of expediency and cost, the new location of the office, qua *field* office, had to meet a number of siting criteria, all of them driven by radar considerations. The overarching question was how to best strike a balance between a maximum of radar coverage and a minimum of signal contamination, or "ground clutter." Yet, underscoring the sociopolitical character of NWS meteorology, the coverage of the would-be Neborough radar was evaluated based on its position not only relative to the broader NWS radar network but relative to Metrocity as well. In fact, whereas most NWS offices are strategically located west or southwest of their respective metropolis given the normal west to east flow of weather, the strong Atlantic influence on the Neborough climate militated in favor of an eastward positioning instead.

Ultimately, after considerable back and forth, X-town was selected, located southeast of Metrocity and also east of the next major urban center. Cheap and readily available property in the form of a developing industrial park satisfied the necessary radar siting requirements. Equally importantly, the X-town authorities were quite enthusiastic about the idea of housing the local NWS office—which, as I was told, they hoped would put their town on the map—and willing to speed up the process. It was not to be a quick and easy transition, however. As has also been the case with a number of other forecast offices around the country, there was another major relocation issue to contend with, and that was employee resistance to the move. Three days before the scheduled moving date, the Neborough office was put under a preliminary court injunction, and the move was delayed for several months. In the end, two members of the staff took early retirement, another one resigned, and two more transferred to a different office as a direct result of the relocation.

Building Tour

Consistent with the rationale behind the relocation to X-town, the exact location of the Neborough office inside the industrial park of X-town was determined by radar siting concerns as well. Towering at ninety feet—at the highest point of the industrial park, where the water tower of X-town used to be—the radar dome overlooks an one-story red brick building across the street, replete with the obligatory American flag and the NOAA and NWS insignia. Just in case there is any remaining doubt, the building's parking lot is flanked by a series of meteorological instruments and satellite antennas.[1]

The structure and layout of the Neborough office follow the standard design of NWS offices built since the modernization of the early nineties. Visitors enter freely through a glass door into a foyer where they may inspect office awards and medals, choose from an array of informational material on weather safety, and read the current seven-day forecast scrolling on a monitor situated behind the glass walls that separate the foyer from the main hall. But here the Neborough office as a government building open to the public ends and the Neborough office as a restricted science area begins. To enter into the main hall, one has to punch in a personal code or be buzzed in by the administrative assistant whose desk overlooks the glass partition. Effectively, these glass doors, while underscoring the epistemic and social fluidity of NWS forecasting, nonetheless separate insiders from outsiders, experts from laypersons, forecasters from the general public. The former are secured in-sight into the future of the weather by virtue of their privileged access to meteorological instruments and data, while the latter are treated to a spectacle of what NWS scientists do with taxpayers' dollars.

Once inside, the hall gives way to three administrative offices, the conference room, and the main corridor. To the right, the corridor leads to the sanctum sanctorum of every NWS forecast office: the operations deck. Upon entering the operations deck for the first time, one cannot help but be impressed by how much it feels exactly like entering an operations deck—a big open space dominated by computer screens, assorted communication devices, map-covered walls. Forecasting life is organized around seven workstations of which three are operational around the clock: the short-term desk, the long-term desk, and the HMT (Hydrometeorological Technician) desk. Forecasting duties are thus routinely split between the short-term forecaster, who, depending on the weather, is responsible for the first twenty-four to thirty-six hours of the forecast, and the long-term forecaster, responsible for the seven-day forecast minus the first twenty-four to thirty-six hours. NWS forecasters will work both desks on a given shift rotation, usually following the weather as they do so by starting out at the long-term desk and ending up at the short-term desk. HMTs, meanwhile, are primarily responsible for collecting and soliciting weather observations from citizen volunteers.

Regardless of their designation, all workstations consist of the same five computer screens, arranged in an L-shape. The outermost left screen provides access to the Internet and the weather data floating therein. Next comes the "text screen," where forecasters receive alerts and type out forecasts and warnings. The remaining three screens, sharing the same mouse and keyboard, serve as the "graphics screens." It is to these last that forecasters give most of their attention—to look at weather information of course, but also to put together digitally the NWS forecast. Regardless of their designation, furthermore, all workstations have access to the same identical data, the high redundancy of information meant to promote interdependence and teamwork among the staff. Weather forecasting, at the NWS and elsewhere, notoriously was one of the first professions to rely systematically on decision support systems to buttress its knowledge claims. It is largely because weather forecasters make aided decisions that they are such high-performing decision makers (Shanteau 2001, 236). To empirically and analytically account for decision making at the NWS, therefore, one must adopt a perspective that recasts cognition as the skillful coupling of neural, bodily, and external resources (cf. Clark 1998). Meteorological cognition materializes at the intersection of people, things, and nature—the result of a complex coordination of a great variety of information systems by means of shared norms and procedures. At the NWS certainly, the institutional skepticism as to whether decision making should—or indeed can—be a process that begins and ends in the skull is quite palpable, with the hub of cognitive activity prominently reassigned to the forecasting workstation instead. Both in practice and by design, meteorological cognition at the NWS is eminently "distributed" (Hutchins 1995a), "extended" (Clark and Chalmers 1998) across physical, virtual, and interpersonal space. If one is to capture the process of meteorological decision making on its own terms, it is thus the task at hand that recommends itself as the proper unit of analysis.

The same organizing principle of distributed cognition is echoed in the ecology of the Neborough operations deck itself, consciously designed to give forecasters optimal visual access to each other's graphics screens as well as to the multiple communication tools that facilitate collaboration across neighboring offices and beyond. When I returned in 2008, the layout of the operations deck had been reconfigured, now anchored by the brand new Situational Awareness Display. Whereas before, one 24-inch screen hanging above forecasters' heads would silently transmit the Weather Channel, the Situational Awareness Display features six 52-inch LCD screens transmitting the feed from the Weather Channel but also from local newscasts, area webcams, radar, and satellite. This reconfigured layout only serves to underscore and magnify the organizing principle of distributed cognition, with workstations and forecasters optimally placed with relation to the Display as well as each other.

Peter (long-term desk) had finished his forecast and was about to get up from his chair when I saw him abruptly sit down again, shoot a quick glance over to his right at the screens of the short-term desk where Norm was busily working, then pull up the Graphical Forecast Editor once again, go into the Temperature grid, and sample with his mouse the forecast values for this evening. He announces to the room, "It will be a chilly night tonight! That snowpack is still holding strong." Norm turns around: "Snowpack?" Peter points up to the Display screens: "I woke up and realized that the snowpack will make the temps quite a bit lower than [model] guidance." Norm turns to look at the Display hanging in front of them. There, in the middle screen, the white snowpack is readily visible in the time-looping composite imagery of radar and current obs[ervations] data. Norm: "Look at that! Thanks for the wake-up call, Petey!" Both men chuckle good-humoredly. I notice that the same composite imagery is also looping in one corner of Norm's left screen. But the snowpack is too small to see there.

Other features of the operations deck include a ham radio station and an NWS radio station, the latter close to the HMT desk because its operation falls under the HMT duties. Beyond the obligatory phone at each workstation, a table behind the short-term desk sports a fax machine and several hotline phones reserved for coordination with emergency management as well as the Storm Prediction Center, the Hurricane Center, and the Ocean Prediction Center. This table is buzzing with activity during severe weather days but typically sits quiet and tidy, awkwardly in suspense. Up above, a monitor streams a live feed from the closed-circuit cameras outside, strategically placed for security purposes but routinely used as weather cameras during a forecast shift.

At the far end of the operations deck, along the long wall, a double line of cubicles serves as office space for the staff. If they are not on deck, this is where forecasters usually spend their shift—following up on administrative work, conducting research, or enjoying some downtime between forecasting bouts. Besides the obligatory mountains of paperwork, cubicles feature photographs and other mementos and generally reflect the personality of their occupants. No such personal touches for the operations deck itself other than three old baseball bobbleheads, which, having survived the office move and their previous owner, have become a sort of mascot for the short-term desk. Even during the informality of weekend and midnight shifts, the Neborough operations deck manages to preserve its all weather all the time intensity and single-mindedness.

To the right, the corridor—now part of the operations deck—continues past the conference room and three offices to a glass exit door and the kitchen, which, with the sole exception of an actual kitchen stove, boasts of every appliance that would make life in a 24/7 workplace livable. To the left, the corridor leads back to the main hall, the mailroom, the bathrooms, the storage room, the electronics technicians' office, and another glass exit door.[2] This does not quite conclude the tour of the Neborough office, however. The careful reader will have noticed that something is still missing. Absent is the server room, housing the computer and other electronic equipment that sustain life on the operations deck. Interestingly, the server room is not accessible through the main corridor but only through the electronics technicians' office, or else through the operations deck. But it is on very rare occasions that a forecaster will enter the server room. Much like the storage room or indeed the electronics technicians' office itself, it is treated as a "backroom," and forecasters will not enter its premises unless they absolutely have to. The physical invisibility of technical work, in turn, promotes and sustains its social invisibility.[3] And while the social invisibility of technical support staff can only receive a passing nod here given the scope of the discussion, the delicate power balance in forecaster/technician relationships deserves underscoring: on the one hand, Neborough forecasters are quite conscious of their absolute dependence on electronic technicians and elaborately careful not to offend or otherwise alienate them; on the other, electronic technicians bask in the knowledge of their power but must suffer the indignity of receiving little recognition—let alone formal awards, as is customary for forecasters—after a successfully handled weather event.

"Team Neborough"

As is typical of all NWS forecast offices, the Neborough office has a total of twenty-five employees, including management and office support.[4] The forecasting staff consists of five Senior Forecasters, five General Forecasters, and a fluctuating number of HMTs and Meteorologist Interns. Crucially, one cannot be promoted to the position of Senior Forecaster without having first served as a General Forecaster, and one cannot be promoted to the position of General Forecaster without having first served as an Intern. To become a NWS forecaster effectively means to be inducted into forecasting the NWS way. This culture of apprenticeship runs deep at the NWS. Not coincidentally, General Forecasters were called "Journeymen" until very recently. Much like physicians, engineers, and other service professionals, NWS forecasters are expected to acquire their training and skills on the job once they have completed a bachelor's in meteorology.[5]

Demographically, the Neborough office again looks very similar to other NWS forecast offices: several of its forecasters are in their late thirties and early forties, married with kids; most are male; all are white. During my stay at Neborough, in fact, there were only five female employees in the entire office, including one senior and two general forecasters. This female to male ratio was rather ahead of the curve, however. Weather forecasting remains an overwhelmingly white, male-dominated field, especially at the NWS. While overt sexism did not seem to be an issue at Neborough, the gender question certainly preoccupied its forecasters, invariably surfacing whenever the topic of discussion would turn to episodes of tension on the operations deck involving females and certain older male colleagues. Such tensions all forecasters understood to be symptomatic of a larger tension between the traditional view, according to which weather forecasting takes place in an agonistic, testosterone-driven, male-bonding field, and the contemporary demands for a more "feminine" environment, which emphasizes collaboration and participatory decision making. It is impossible to know in what ways and to what extent my presence as a then 33-year-old female sociologist, with what were described to me as "feminist views," may have "contaminated" the field. Still, the fact that I felt so welcomed and at ease during midnight shifts, presumably considered "male time" in other offices (cf. Fine 2007, 39), is a testament to the kind of gender work accomplished on the coed operations deck of the Neborough office. Even if the persisting homosociality of the weather forecasting world is slowly being eroded in places like Neborough, however, this is still early days for women forecasters at the NWS. To be a good NWS forecaster still very much implies proving you are the right *man* for the job. To enter and stay in the profession, women must be determined to go to extra lengths to "pass" as forecasters.[6] Characteristically, it was the men at Neborough who would complain about the grueling rotating shiftwork, never the women. And it was the women who had felt the need to add two chairs in the female bathroom and generally transform it into a sort of female space/time off, a "safe zone," where they would frequently congregate to confide, laugh, and cry.[7]

If gender was not an overt source of tension, tension nonetheless abounded on the Neborough operations deck. Early on, while still in the process of discussing the parameters of my upcoming visit, I was warned by the office director that this was a place with "strong personalities" and that I should prepare myself for the eventuality that some forecasters might not be amenable to my presence on deck during their shift. Months after my arrival at Neborough, various members of the staff would still inquire, half-jokingly, whether the reason I specifically chose their office was because it was infamous for its "challenging work environment." And I was advised by a forecaster at a neighboring office that "Every office has its problems, but these guys have it magnified." My visit to Eastern Region Headquarters would only further confirm Neborough's notoriety. Officials there alluded to its "personality range that is wider than usual," or to its having "a more challenging mix than elsewhere." Although my experience at Neborough was nothing but positive, I did come to understand why the office director thought it wise to forewarn me. Things had apparently been "much worse ten, fifteen years ago [because] some of the 'problem people' left the Service when the office moved." Still, strained incidents among the staff persisted. In spite of conflict-resolution training sessions and administrative efforts toward team building, the overwhelming consensus back in 2004 was that only when certain people retired "will this black cloud truly be lifted."[8]

In this light, the salience of the phrase "Team Neborough," meant to include all office employees but primarily applied to the forecasting staff, acquires special significance. I have heard forecasters use it ironically when discussing the latest office drama. I have heard them use it proudly after a successful forecast. And I have heard them use it excitedly as they prepare to face a nasty weather system. It was most likely coined in the early nineties when, as part of a broader organizational effort to institutionalize collaboration during the production of the forecast, the term *forecast team* first made its appearance in the official NWS vernacular. It took roots in the Neborough environment because it encapsulated all that is exasperating and inspiring about the place and about NWS forecasting. It resonated of "war stories," of men working as one in the face of weather danger after nearly losing their way in the heat of the fight. It represented an ongoing identity struggle—both a predicament and a promise. And, in an office full of sports fans, it certainly aimed to inspire a sort of Bourdieuan association between team spirit and scientific identity (cf. Lenoir 2006, 26).

Yet one would be too quick to dismiss Neborough as dysfunctional or extreme—the exception to a rather cohesive and homogeneous office fleet. On the contrary, given the organizationally rather flat field structure of NWS forecasting operations, one can reasonably assume that there is considerable variation in how NWS offices may acceptably articulate who they are and what they do. If the bird's-eye perspective from Eastern Region Headquarters helps identify what stands out as unique about the Neborough office, Fine's ethnography of three midwestern offices illuminates the relative distinctiveness of office life across the NWS.[9] Indeed, a comparison with the Neborough case gives a fair sense of just how considerable the range of variation among office idiocultures can be. Most striking in this respect are the differences between offices that, at least on paper, would appear to be quite similar to each other. The Chicago office and the Neborough office were both established back in 1870 in key metropolitan areas under the Army Signal Corps. Both have a director who is averse to micromanaging, and both draw their strength from a long and distinguished meteorological heritage. Just as importantly, both have had a reputation for being "problem offices," compelling NWS leadership to bring in new blood in order to "shake things up" (cf. Fine 2006, 5). And yet, whereas the Chicago office has a "tradition of tradition" (Fine 2006, 5) and prides itself on resisting authority and change, the Neborough office laments being surrounded by "dinosaur" offices. Whereas Chicago forecasters do not engage in research and in fact explicitly do not see research as part of their job description (Fine 2007, 61ff.), nearly every forecaster at Neborough maintains an active research agenda, presenting at professional conferences and, at times, even publishing in peer-reviewed journals. Whereas Chicago would appear to be characterized by strong interpersonal bonds and solidarity, complete with nicknames, running jokes, a bowling team, and other outside of work socializing, socializing at Neborough is more reserved and fragmented, albeit no less genuine and heartfelt. And neither does Neborough's identity resemble the idiocultures of the other two offices Fine observed. The Flowerland office was established during the NWS modernization of the early nineties and therefore consists of a cohesive cohort of much younger forecasters, (selected to be) in step with the new NWS "party line." The Belvedere office shares some similarities with the Neborough office in that it sees itself as progressive, but otherwise it would appear to be too orderly and regimented by Neborough standards.

While not equally aligned with its bureaucratic logic, all of these offices nevertheless daily fulfill the NWS mission with integrity and professionalism. Fine goes to some lengths to drive this point home about Chicago, Flowerland, and Belvedere after establishing that each one operates according to its own idioculture, and I can most certainly say the same about Neborough. Just as it is important, then, to remain mindful that this analysis of meteorological decision making is emplaced within a particular microcommunity of weather forecasters, it is equally important to bear in mind that these forecasters are simultaneously members of a broader technologically mediated interaction order and a broader subculture spanning 122 such microcommunities, all of whom identify with the NWS mandate and rely on a more or less shared stock of heuristics, techniques, technologies, information resources, and norms to bring this mandate to fruition.[10] It is their membership in the latter cultural group that makes Neborough forecasters the focus of this book.

This section has delved into what makes Neborough distinct in order precisely to highlight what makes it a typical NWS office. In the pages and chapters that follow, I will take care to clearly demarcate observations that I am not confident extend beyond the Neborough experience. Differently put, it is my contention that, on the whole, my observations—including the embedded field note and interview excerpts—about meteorological decision making at the Neborough office reflect forecasting practice at the NWS at large, or else they have not been included in this analysis. On this note, it is finally time to take a first proper look at the day-to-day process of generating the NWS forecast.

Producing the Weather Forecast

Protecting America's life and property against the calamities of the weather is a daunting task to manage, above and beyond the formidable meteorological challenges involved. There is a tremendous range of "weather" for the NWS to keep an eye on: land weather, airport weather, marine weather, fire weather, hydrologic weather. To properly protect America's

CHAPTER TWO

life and property against the calamities of the weather, therefore, NWS forecast offices are operational around the clock. "The weather never sleeps and neither do we" is the usual stock phrase Neborough forecasters use to enlighten outsiders about their schedule and, by extension, their importance.

The primary responsibility of an NWS forecast office, of course, is to advise and alert about the potential for hazardous weather. Only when hazardous weather warning requirements have been met do forecasters turn their attention to routine products and services. Indeed, during a hazardous weather event, the NWS becomes transformed into what Fine (2007, 40) calls an "activated organization . . . verging on being overwhelmed and understaffed, until routine can again be established." In this, NWS forecasters readily resemble firefighters, paramedics, and other first responders—primed for hazardous weather, they seem perpetually caught in a lull before the storm. But pushing the analogy any further may be misleading. In contrast to firefighters (cf. Desmond 2007, 81ff.) and other emergency professionals, NWS forecasters experience little or no workload downtime.[11] In 2008, five years after the implementation of the IFPS, with the work schedule radically revamped to accommodate for the realities of the new forecasting process and the new forecasting process already an old routine, Neborough forecasters were still struggling to keep up with the weather. To be sure, their workload has been exacerbated by an expanding list of IFPS-driven forecast responsibilities, which, coupled with a series of staff cutbacks due to a shrinking budget, force offices to "do more with less," to use another favorite NWS catch phrase. What drives this heavy workload in the first place, however, is the fact that we have come to recognize "the weather" not as a finite phenomenon, like a wildfire or a robbery, but as ever present and relevant. As competition among the various weather forecast providers emboldens our appetite for more, faster, better weather information, what counts as the weather further expands in detail and significance. And so, too, does the charge of NWS forecasters. NWS forecast offices more closely resemble a newsroom in this respect—busy during routine operations, verging on understaffed during emergency conditions.

It is this trait of the weather that, in the context of NWS operations, makes weather forecasting an auspicious case study for probing the process of decision making in multiple decision-making regimes. But first, the basics. By way of an introduction to the process of meteorological decision making, the remainder of this chapter goes over the main components of a typical shift at the Neborough office: data analysis; deliberation;

and finally, the actual *doing* of the forecast. In practice, of course, these components are thoroughly intertwined. There is no actual moment when diagnosis ends and prognosis begins. Rather than constituting the means and ends, respectively, of forecasting action, diagnosis and prognosis are in fact "two names for the same reality" (Dewey 1922, 36). Neborough forecasters never switch from a diagnostic to a prognostic frame of mind—they just continue making increasingly more consequential decisions as the forecast submission deadline draws nearer. If anything, meteorological prognosis analytically *precedes* diagnosis, as will become evident in the following pages. This empirical reality of NWS forecasting, while unintelligible from a rational choice perspective, is entirely consistent with pragmatist accounts of the decision-making process. "We do not use the present to control the future," writes Dewey (1922, 322); "We use the foresight of the future to refine and expand present activity." The formal distillation and ordering of the NWS forecasting routine into a diagnostic, a deliberative, and a prognostic component denotes therefore the temporal, or processual, unfolding of forecasting action rather than its analytic structure.

Taking Over the Hot Seat

No shift can start without a briefing by the outgoing forecaster to get the incoming forecaster up to speed with the big weather picture and developing concerns. As in any other work setting whose rhythm is dictated by a shift schedule, weather briefings form an essential, organic part of the forecasting routine as they allow for efficient resource management, minimize duplication of effort, and promote forecast-to-forecast continuity. Weather briefings almost always occur right at the workstation of the outgoing forecaster. Indeed, if the incoming forecaster does not find the outgoing forecaster at his desk as she walks in, she will seek him out in his cubicle and, together, they will walk back to the operations deck to begin the briefing. This is not a mere formality but a testament to the role of *screenwork*—that is, the processing of information via computer screens—as the organizing principle of meteorological expertise (Daipha 2013). As already noted, if they cannot see it, forecasters cannot think, never mind talk, weather. The departing forecaster relies on the computer screens to make a case for his forecasting decisions, reasoning through the assortment of weather displays he flags as pertinent. For her part, the incoming forecaster relies on the computer screens to keep up with the action, to bring into focus what portends to be the weather forecasting problem of the day. As will become apparent time and time again, screen-

CHAPTER TWO

work forms the backbone of every aspect of the meteorological decision-making task.

A briefing is in reality two briefings in one, a briefing about forecast concerns and a briefing about technology malfunctions, and forecasters at Neborough will invariably cue "weatherwise" and "equipmentwise" to signal the end of the initial pleasantries and the start of the briefing or to segue into the next section. Depending on the weather situation du jour and the familiarity of the incoming forecaster with the current weather system—in other words, depending on whether she has worked that desk the previous day—briefings may last from several seconds to over ten minutes, not infrequently turning into protracted meteorological discussions on model performance and biases or similar past weather events. Throughout, the two forecasters will be poring over the computer screens, the outgoing forecaster guiding the action with the computer mouse. Anything and everything deemed relevant information can be included in the weather briefing: model(s) of choice and reasoning behind it, forecast dilemmas and ultimate decisions, remarkable personal weather observations and puzzling spotter reports, verification concerns, deliberations with neighboring offices, weather features to watch out for, upcoming hazards and how they have been addressed so far, interoffice coordination issues. The conversation is relaxed and collegial, a good-natured back and forth. Yet, even between forecasters who know each other well and have a high regard for each other's forecasting skill, the exchange is clearly underwritten by a handing off the baton dynamic: the outgoing forecaster is eager to make his forecast stick, especially if he is coming back in a few hours, while the incoming forecaster is intent on not missing a beat but not necessarily committed to the particulars of the existing forecast.

Margaret (short-term desk, day shift): So you guys still thinking some action today?
Dick (short-term desk, midnight shift): Yes. Yes. Todd [at the neighboring office to the southwest] and I were talking it over, and there's just enough cold air advection aloft, and it's a sharp enough upper trough, that I was not going to go against it. . . . But, as you can see here, there's a fair amount of action upstream right now in [adjacent states to the southwest], and the potential is there for all that to work in and just stabilize the air mass out of all severe possibility. So, it's not something I'm entirely confident in, but with the other factors in place I just showed you . . .
Margaret: So, it must be all instability aloft, because it feels pretty comfortable out there.
Dick: Yeah. . . . I mean, see here, we are eventually looking at a minus twelve [degrees Celsius] at 500 [millibar atmospheric pressure level]. But it's going to be this

evening before it gets down into our part of the Northeast. . . . So, that's part of the problem, too, that . . . if anything busts the forecast is that the cold pool . . .

Margaret: Takes forever?

Dick: . . . takes forever to get in. So that's something to keep an eye on. . . . Anyway, I did what I could; it's in your hands now.

"Equipmentwise," the briefing can take up an equal amount of time—hardly surprising given the big science character of NWS forecasting operations. Technology fails forecasters in big and small ways, from the radar going down during a severe storm to the server being slower than usual, and it fails them in multiple ways at once. The log of any given shift during my stay at Neborough contained at least two equipment malfunction entries, with that number doubling during hazardous weather conditions. The weather spares no one, certainly not the people tasked with anticipating its every move.

Finally, a "changing of the guard" of sorts takes place. Throughout the briefing, the outgoing forecaster has remained seated at the desk chair with the incoming forecaster standing or leaning against the desk next to him. The briefing completed, the outgoing forecaster will now stand up and, sometimes with some kind of verbal or nonverbal flourish, he will offer the seat to his relief of the day.

The incoming forecaster is now in charge of the workstation. But not until she has adjusted the computer screens according to the settings stored under her user profile will she have truly taken control of the desk and of the weather. The customized, unique combination of weather display formats, color graphics, and sound alarms effectively transforms the workstation into one's personal workspace, and Neborough forecasters are quick to switch over to their profile the moment they claim the seat. That is especially so because, stored under a forecaster's name but accessible by all, is a set of personal "best practices" for looking at weather information: the so-called *Procedures*, accumulated over one's career and updated as necessary. To study the wind along the atmospheric column, for example, one forecaster might prefer the 850, the 500, the 250, and the surface millibar height charts, color coded just so, while another might routinely find the 700, the 500, the 200, and the surface charts more insightful. Where the Procedures become truly useful, however, is in their ability to recall elaborate composites of data graphics in a matter of seconds. As a result, the variation among forecaster profiles can appear quite staggering. Yet, despite their seemingly idiosyncratic nature, forecasters' profiles reflect eminently social decisions, the result of apprenticing at particular meteorology programs and forecast offices, under particular mentors, with particular technologies, and so on (see Daipha 2010). For example, unlike other Neborough forecasters, Biff and Phil have primarily organized their Procedures according to weather scenarios, something they learned to do at X-University, which they attended several years apart. And Margaret and Phil are in the habit of looking at weather data in really busy four-panel displays, something they picked up, as it turns out, while interning in the same forecast office in the Midwest.

To be sure, our forecaster does not, as a rule, *have* to change the previous weather display settings. Rarely did I witness a Neborough forecaster request that the display settings be changed or explained when hunched over someone else's workstation—following the weather was straightforward enough. In fact, there is an argument to be made that no profile presets are truly necessary, and that the default settings would more than suffice for the task ahead. Certainly, most forecasters like to claim that relying on their Procedures is only an issue of expediency. In practice, however, temporal constraints and the threat of information overload lead to a near absolute dependency on such preset templates for studying the weather.

2:45 p.m. Phil was working an administrative shift in his cubicle today but has been called on forecast duty because of the potential for severe weather later this afternoon. He logs into one of the vacant workstations, turns to the left graphics screen, goes to File/Procedures/Select User ID, selects his user name, clicks on Severe Weather Tools from the drop-down menu, and selects to load all six of the included information sources from the new drop-down menu. Repeats the same process for the middle graphics screen, this time clicking on Meso[analysis] Stuff and selecting five information sources (mostly guidance products from the Storm Prediction Center) out of approximately twenty-five. While waiting for the data to load, he next turns to the right graphics screen but now selects Biff's user name and clicks on Severe_Neborough Radar_Right. He tells me he has been meaning to copy this procedure into his own user profile. He likes how "nice and clean" Biff has set up his radar info for right-moving storms. . . . Within fifteen minutes, in consultation with Tom (short-term desk), Phil is ready to press "Send" for the first severe thunderstorm watch of the day. He rubs his hands together in excitement: "And we're rolling!"

In effect, these procedural shortcuts do not simply represent a handy meteorological tool kit of sorts but the very embodiment of each forecaster's skill set, there for the entire organization to use and leverage. More precisely, these repertoires of meteorological perception are technological externalizations of the cognitive schemata governing forecasting practice at the NWS as already distilled in particular techniques and know-how. As such, they are never just symbolic or merely aesthetic but have instrumental value as well. Much like a writer's attachment to a particular word-processing program, a particular font, or a particular zoom level, such seemingly trivial choices become important for getting into "flow" (Csikszentmihalyi 1990). In a profession that claims to be as much an art as it is a science, personalized forecasting routines, if not expressly encouraged, are nonetheless considered part of the process. And this is precisely why the exclamation "I was wondering why the weather is all off today!," uttered by a forecaster upon realizing he had neglected to switch display profiles, does not ring false on the Neborough operations deck.

Looking at Weather Data

The incoming forecaster continues immersing herself in the weather forecasting problem of the day—*continues* being the operative word. Even if this were to be her first day back after a vacation hiatus, by the time the briefing is over the incoming forecaster has already formed an expectation about how the atmosphere is going to behave. This initial impression will inevitably become adjusted over the course of the shift as she is exposed to more and new information. Importantly, however, the briefing sets the tone for the entire shift. For it brings into view for the first time what are framed as the forecasting challenges and aims of the next few hours. These aims, or "ends-in-view" (Dewey 1922, 36, 223ff.), are still quite hazy, of course—indeed, they will not properly crystallize until after the fact—but they are bound to exert an inordinate influence on the decision-making process of the incoming forecaster.

The tendency to disproportionately rely on the first available information cue when making a decision under conditions of uncertainty is known in cognitive psychology as the "anchoring heuristic" (Tversky and Kahneman 1974). Simon's theory of bounded rationality has inspired a long tradition of psychological studies on the satisficing rules of thumb people employ to solve problems when faced with imperfect information. According to this line of research, by reducing task complexity, anchoring and other satisficing heuristics make decision making more tractable and, indeed, often produce approximately correct judgments. Because they rest on inherently flawed logic, however, they introduce bias into one's calculations and can lead to systematic and serious errors in judgment. Hence the self-professed goal of scholars in the "heuristics and biases" tradition to analytically concentrate on instances where satisficing fails us (e.g., Kahneman, Slovic, and Tversky 1982).

Alas, there is no neutral, objective position from which to assess information. In practice, all decision making is contextual and perspectival, therefore "biased," and it would be absurd and reckless to expect it to be otherwise. Consequently, the NWS seeks very deliberately to anchor the reasoning and practice of its forecasters into its own standards for evaluating and acting on weather information. The "handoff" between forecasters during shift briefings represents in this respect a crucial organizational mechanism for biasing decision making toward a consistent and harmonious NWS forecast. Even so, it is only one part of a broader effort to cultivate a particular aesthetic of meteorology in NWS forecasters so as to ensure they possess the right practical sensibility. This sensibility is what Plato and Aristotle refer to as *aesthesis*: a situational awareness that compels decision makers toward the institutionally sanctioned course of action while also insulating them from the attractions of competing alternatives. Buttressed by the strong ethos of apprenticeship at the NWS, the inculcation of the proper meteorological aesthesis occurs quite deliberately in a wide variety of settings, from required refresher teletraining courses and NWS Directives to office workshops and Senior Forecaster–General Forecaster interactions. And it is continually rehearsed in the course of forecasters' daily practice of looking at and appraising weather data.

Looking at weather data is a truly elaborate and plodding process, not least because of the sheer amount of information one has to absorb, and it is not unusual for Neborough forecasters to scribble notes and comments on a nearby pad of paper as they do so. Indeed, the beginning of the shift is one of the very few quiet times on the operations deck. Our forecaster spends several minutes at a time literally staring at one of the graphics screens, the only sound in the room being the clicking of the mouse as she eventually moves on to the next weather display. What to the uninitiated comes across as a dauntingly complicated two-dimensional map riddled with weather symbols transforms itself into a four-dimensional view of the atmosphere under a forecaster's knowing gaze. Because, of course, forecasters are expected to somehow propel themselves out of the loop of current weather conditions forever pulsating on their screens and into the future. They must see beyond the three-dimensional interaction of weather elements in front of them and behold them forward in time, in four dimensions.

To foresee the weather, forecasters draw on their abilities at visual pattern recognition, a skill set frequently invoked and highly valued across the profession. Granted, meteorological pattern recognition relies heavily on computer-generated forecasts, which, as the shift progresses, come to overtake observation data on forecasters' screens. For the moment, however, all our forecaster has eyes for are visualizations of current weather conditions. Only after she has familiarized herself with these data and settled on a general course of action will she begin to examine computer model data. At the Neborough office at least, the credo that model forecasts are "just guidance" runs deep, and the one or two forecasters reputed to consistently stray from that path and treat model forecasts as the final word are held in low regard. The quest for autonomy and control over nature and machine is a highly resonant idea in the world of weather forecasting. To be professionally acknowledged as a forecaster, one is expected to counter one's own skills at pattern recognition against those of the computer's. And so, the algorithmic brute force of model forecasts poignantly, if temporarily, absent from her range of vision, our forecaster is intently looking at observation data, straining to sort it out into a pattern, to *see* the weather contained therein so as to project it into the future.

She looks at visible, infrared, and water vapor satellite imageries and notes interesting "features of the day." She looks at upper-level pressure charts, derived from weather balloon soundings, and notes the positions of troughs, ridges, and jet streams relative to the key features identified on satellite. She looks at lower-level pressure charts and notes the position of cyclones, anticyclones, fronts, troughs, and dry lines, linking these features to the upper-level conditions. She looks at surface weather maps, derived from automated observing stations, and locates centers of low pressure, fronts, areas of precipitation, and squall lines, trying to relate changes at the surface level back to the larger-scale atmospheric conditions.

This visual appraisal of weather data is long and thorough, but one should not mistake it for exhaustive or evenhanded. Our forecaster is working her screens not simply with the abstract goal of forecasting the weather but with a concrete, albeit fuzzy, objective in view: to solve the weather problem of the day as provisionally established during the shift briefing earlier. This provisional target becomes part of the means through which she now charts a diagnostic course of action, sorting through the welter of available materials, prioritizing among competing information sources, probing and interrogating the relevant data (cf. Dewey 1922, 226; Whitford 2002, 37–38). It is, then, this provisionally identified weather problem that, by foregrounding certain lines of inquiry and backgrounding many others, makes the task of diagnosing the weather problem of the day orderly, systematic, and thorough. One must first settle on a certain goal in order to have something to work *with*, not just toward. To do otherwise would not amount to more rigorous or objective judgments but to blind judgments, an entirely impractical and preposterous proposition.

But the shift briefing is only the first, albeit crucial, input our forecaster will receive as she forms an idea of, and a solution to, the weather problem at hand. In the course of her diagnosis, she will invariably adapt her objective, as originally conceived, to ongoing developments: her diagnostic findings, unexpected reports, evolving weather events, deliberations with colleagues, guidance from forecast models, the new run of the forecast models, and so on. The changes can range from relatively straightforward fine-tuning during fair-weather shifts to dramatic flip-flopping during an outburst of hazardous weather. Importantly, this progressive redefinition of the situation is shaped by material interactions just as much as it is shaped by social interactions. In fact, although I make a point of drawing attention to the materiality of meteorological judgment and decision making below, keep in mind that any distinction between the material and the social is entirely analytical because, in practice, they are inseparable elements of proper screenwork.

Despite the variation in the Procedures they use to screen weather data, forecasters follow the same general path as they familiarize themselves with the forecasting problem of the day. Our forecaster will thus start at the hemispheric scale and work her way down to the national or synoptic scale, the regional- or mesoscale, and the local or microscale. This diagnostic sequence corresponds to the so-called forecast funnel (Snellman 1982) professionally considered so fundamental and emblematic of the weather forecasting process that Neborough forecasters always reserve a PowerPoint slide for it in their citizen volunteer training sessions. How far downscale our forecaster goes will depend on which workstation she happens to be sitting at that day. A long-term forecaster will concentrate on hemispheric and synoptic weather patterns by examining satellite imageries, atmospheric pressure charts, and other relevant surface and upper air analyses, whereas the short-term forecaster will quickly move on to focus on the smaller-scale features of the Neborough region, and she will spend considerable time "dissecting" (cf. Knorr Cetina and Amann 1990) current local weather conditions derived from automated weather stations, balloon soundings, satellite, and radar.

In their quest for ground truth, forecasters endeavor of course to maximally exploit the capabilities of their screens. The aim is to take all of the weather in at once, to embrace it *synoptically* so as to truly see it (cf. Latour 1999, 51). In effect, meteorological perception exists in a state of perpetual motion between a macroscopic and a microscopic orientation toward weather information. Neborough forecasters thus keep switching between the zoom in/out function as they slide down the forecast funnel. And rarely do they study the weather in the form of static two-dimensional maps. They prefer to behold it in a 32-frame animated loop instead, momentarily freezing it to examine it in detail, then bringing it back to life again. More often than not, in fact, looping graphics are superimposed to form a variety of composite images and/or they are juxtaposed in four-panel displays, only to be replaced by different graphics and different composites as forecasters proceed with their visual hermeneutics of the weather. Because the atmosphere cannot be made into specimens and brought inside the controlled environment of the operations deck, forecasters work hard through their screens to channel and compile information into an entity in its own right: the weather-on-screen. Thus "appresented" (cf. Knorr Cetina and Bruegger 2002), the atmosphere becomes all but reconstituted into something calculable:[12] dynamic yet contained, complex yet refined, chaotic yet predictable.

With a fair understanding of the weather forecasting problem of the day and a rough plan of attack, our forecaster now turns to look at computer model data. Depending on her workstation and, therefore, how far down the forecast funnel she has to go, this change of pace usually occurs twenty minutes to an hour into the shift. At her disposal are a host of forecast models. Some are operated by the NWS, others by U.S. universities or other research centers, still others by weather agencies in Canada, the European Union, or Great Britain. Some are global models, providing predictions for the big weather trends around the world every twelve hours out to sixteen days. Others are regional models, focusing on a specific part of the world, such as North America, out to two or three days but of considerably higher spatiotemporal resolution and updated every three or six hours. Still others are hourly updated short-range models serving specialized needs, such as air traffic control, severe weather forecasting, or energy management. Indeed, even after accounting for the differential model guidance requirements of each workstation, there are still so many models and so much model data at hand that they are almost too much. It was not unusual for Neborough forecasters to complain of information overload in the same breath that they boasted of the wealth of meteorological guidance at their disposal.

As is the case with observation data, the decision about which models to look at is motivated by profoundly practical considerations. During the first round of my fieldwork, all Neborough forecasters would thus closely study two forecast models: the GFS (Global Forecast System, formerly known as Aviation/MRF) and the NAM (North American Mesoscale, formerly known as ETA). The reason is not merely because these are domestic models or even because they are operated by the NWS. Rather, at the time, these were the only models whose graphics were available to forecasters in gridded form. These, in other words, were the only models forecasters could use to directly populate the forecast grids in the IFPS Graphical Forecast Editor. As a result, all forecasting deliberations essentially revolved around whether to align oneself with either the GFS or the NAM camp. The remaining models, meanwhile, served to tip the decision-making scale one way or the other, and they occasionally acted as "fillers" whenever forecasters decided they had to adjust by hand an entire aspect of the forecast—usually the Sky Cover and the Probability of Precipitation grids—because it was so poorly handled by both the GFS and the NAM. By implication, beyond these two models, there was considerable variation in which other models, if any, forecasters would consult during a given shift. Most would at least take a look at the ECMWF (European Center for Medium-Range Weather Forecasts) model, some might also look at the British UKMET model or the now-discontinued NGM (Nested Grid Model), but a fair number would not even consider the Canadian model because, as they acknowledged, it was not delivered directly to their graphics screens via AWIPS, and they would have to look it up on the Internet. For the same reason, very few Neborough forecasters consistently paid attention to the RUC (Rapid Update Cycle) model, and only one forecaster would occasionally consult the MM5 (Fifth-Generation Penn State/NCAR Mesoscale Model).

For all intents and purposes, then, NWS forecasters today arrive at a meteorological prognosis by looking at gridded model data. To work toward the abstract goal of making a forecast, they work with the concrete goal of producing the forecast grids. Therefore, it is the practicalities around the production of the grids that help ground and inform their forecasting process and shape their decision making. These practicalities, I should hasten to add, do not take center stage because of the overstretched NWS forecasting schedule or the threat of information overload, although this certainly is how Neborough forecasters typically account for what they consider nonmeteorological choices in their forecasting process. Rather, as already noted, the issue is much more basic than that: it is a matter of managing the task at hand, including managing the time and information involved. Decision making, at its core, is nothing more than sorting through and prioritizing competing alternatives—a fundamentally practical problem requiring a fundamentally practical approach. In this respect, there is no more rational course of action for a forecaster to take than to analytically privilege those models that are most practically suited to the forecasting task at hand—in this case, gridded models. And so, when I returned to Neborough in 2008, despite an admittedly greater information load and an equally heavy schedule, forecasters would not just study the GFS and the NAM anymore, but they would also study the

ECMWF model, and the RUC model, and the SREF (Short Range Ensemble Forecasting) model, and the WRF (Weather Research and Forecast) model, because they all were now available in gridded form on their screens.

Back at the Neborough office, our forecaster has just started looking at model data. But this change of pace does not signal a shift from diagnosis to prognosis. For one, there is no clear break from the one to the other, with prognostic judgments serving to guide and refine diagnostic judgments and vice versa (cf. Dewey 1922, 72). For another, although she has by now formed a relatively clear picture of the weather problem of the day, she still needs to evaluate how the forecast models under consideration are handling the current weather situation in order to figure out to what extent she can rely on their prognostic guidance. Looking at model data during this initial, familiarization phase is thus always performed within the context of observation data. Forecasters will either overlay model data over a weather map of current weather conditions, or they will juxtapose models and observations on two adjacent graphics screens. The rationale behind this diagnostic strategy is twofold: to assess how "naturally" the weather conditions forecast by the model under consideration are evolving from current weather conditions, and to establish how the particulars of the weather system slated to reach Neborough in a few hours are presently verifying against ground truth. The immediate objective is to weigh the available options and decide which model offers the most "reasonable" solution to the weather forecasting problem at hand. Ultimately, of course, NWS forecasters are organizationally expected to improve on a computer-generated forecast. "Beating the models" is one of their most cherished mantras because it is so essential for legitimizing their claim to expertise and for justifying their professional autonomy. First, however, one must decide "which pony to bet on," as a Neborough forecaster liked to say. At this stage of the process, what matters is settling on which model(s) to do the forecast with, not against.

After this preliminary diagnostic assessment, and before a second round of more targeted, prognostic engagement with observational and model data, our forecaster will turn to consult any number of pertinent technical discussions issued by the NWS National Centers for Environmental Prediction, ranging from model diagnostics to hazardous weather outlooks.[13] While typically no longer than ten minutes in duration, this is a no less crucial step in the decision-making process. Forecasters take it for granted in their deliberations with neighboring offices that everyone is sufficiently familiar with the technical discussions of the day. Common reference points of the "official line" on the weather past, present, and future that they are, NWS technical discussions not only forge a consistent synoptic weather picture across the minds of forecasters but also lay the foundation for the forecast of each office, thus mitigating gross disparities among them. This is not to say that forecasters are expected to, or in fact will, adhere to the feedback from the National Centers. Just like forecast models, technical discussions are considered "guidance, not the law," and it is left to the discretion of local forecasters to determine their relevance in field operations. Yet, as our forecaster now turns again to look more closely at the model guidance, straining to chart a concrete prognostic path of action, it would do well to remember that the baseline of the Neborough forecast was already organizationally decided well before she took her seat in front of the screens.

Meteorological Deliberation

Whenever I would ask Neborough forecasters to describe the forecasting process for me, they would invariably preface their response with the equivalent of the phrase, "Everyone does it differently," before launching into a description of *their* process. True enough, our forecaster is sure to prefer particular weather maps to establish the weather problem of the day, favor particular models to populate the IFPS Graphical Forecast Editor with, use particular IFPS software tools to edit the forecast grids, and follow a particular grid sequence while doing so. As already discussed, however, weather forecasting, like all systems of knowledge, is predicated on a communal *aesthesis*: shared modes of looking, reasoning, and doing. For the sake of narrative convenience, I have thus far been concentrating on a single forecaster to analyze the production of the NWS forecast. But of course forecasting the weather at the NWS is never a solitary achievement but a deeply collaborative organizational enterprise. What is more, as I illustrate below, it is an *interactive* organizational enterprise.

Earlier, I claimed that the time of a forecaster's familiarization with the weather problem of the day is one of the quietest on the operations deck. Soon enough, however, after having studied her screens for a while, our forecaster is bound to break the silence and comment on an unusual observation report or the intriguing way a model is handling the approaching weather system. During my initial weeks at the Neborough office, I was constantly mystified by how seemingly random five-word declarations, uttered out of the blue by a forecaster ostensibly deep into his screens, were picked up and answered right on cue in equally long sentences by the proper forecaster, also deep into his screens. Erving Goffman (1981, 134–35) has called such flurries of conversation in the midst of intense work activity an "open state of talk," where the usual conversation rituals become subordinated to accomplishing the task at hand. The unceremoniality of this back and forth was particularly striking at Neborough at the time because forecasters had been paired up with a shift partner and had consequently developed old married couple habits. As I became better versed in the meteorological jargon and the shift routine and got more familiar with the meaning and logic of these syncopated interactions, I came to realize that they were a form of "object-centered sociality" (Knorr Cetina 1997), their content and timing dictated in and through screenwork.

Doug (long-term desk): Wow! The Euro[pean model] is dumping some major rain on us Monday!
Tom (short-term desk) [*without missing a beat*]: Yeah . . . but the GFS brings it more over land, doesn't it, and so it doesn't get the chance to strengthen.

* * *

Wayne (short-term desk) [*muttering to himself*]: Metrocity Airport will be tricky . . .
Phil (long-term desk) [*without turning*]: I say fifty PoP [probability of precipitation], one two [hours] at most, gone by seven. . . . What do you say?
Wayne: I say fifty-four PoP. [*laughing*] Great minds think alike, boy!

Most often, especially on meteorologically quiet days, such exchanges remain brief, fleeting. And they typically start off with a statement, not a question—a forecaster bouncing ideas around. After a couple of back and forths, the discussion usually exhausts itself; but even then, if warranted, the responding forecaster might pull up relevant data on his screens before replying. Remember that thanks to the principle of redundant access to information governing their workstations, Neborough forecasters share a real but also a virtual world. Not only does the responding forecaster not need to get up from his workstation, therefore, but he also does not need to look away from his screens—he has the same data, he is familiar with at least the synoptic weather picture having gone down the same forecast funnel, and he is cognizant of the rhythm of the forecasting routine on the other desk. Indeed, if he does look over to the screens of the other workstation, this is bound to be a more involved exchange. And were he to end up wheeling himself over to the other workstation to further deliberate the issue while looking at the *same* screens with his colleague, this is a strong indication that something extraordinary is going on, something meteorologically complex or plain fascinating. The cognitive and/or emotional weight of the situation is such that a physical sharing of virtuality, a literal putting of heads together, is called for. That is why, upon hearing

CHAPTER TWO

and seeing the huddle, it does not take long for other members of the staff to feel prompted to leave their cubicles and come join the discussion and the excitement on the operations deck.

The prime ground for meteorological deliberation, to be sure, is hazardous weather forecasting. The par excellence case would be a severe weather episode, such as a summer storm, where the need for near-real-time decision making has led to the adoption of a "severe weather team" approach to forecasting (see chap. 5). Nevertheless, the new distribution of routine shift duties along the temporal axis of the forecast constitutes a cognitive division of labor toward a common goal and, as such, is also supremely conducive to instances of collaboration. A considerable amount of these ad hoc meteorological exchanges between the short-term and the long-term forecaster are therefore clustered around the time frame of the split between the two workstations. In contrast, collaboration during benign weather appears to have been nearly absent in the pre-IFPS era (Fine 2007), when shift duties were typically split between the aviation and the land forecast. The IFPS Graphical Forecast Editor only serves to further encourage meteorological deliberation. Upon finishing a forecast grid, our forecaster is apt to let its graphics loop from days one through seven of the forecast to make sure they progress "naturally" and reflect her vision of the future. Were she to notice something is off in the part of the grid that falls under the temporal jurisdiction of the other workstation, she may comment—or directly inquire—about the matter, a comment that often leads to a minor or major adjustment of the given grid by one or both forecasters on shift.

Yet techniques and technologies conducive to collaboration do not by themselves a collaboration make. After all, the concept of the forecast team and the notion that everyone, from Senior Forecaster to HMT, shares ownership of the forecast are relatively recent developments at the NWS. Neborough forecasters still reminisce about their fear and awe when addressing a Senior Forecaster as Interns not so long ago. In those days, Senior Forecasters had final say on the land forecast and often would not let General Forecasters work on it at all, assigning them to the less prestigious aviation or marine forecast instead. It was not until the nineties and the NWS modernization that the notion of weather forecasting as teamwork became a staple in the NWS operational discourse and directives. The final coup came with the reclassification of all Senior Forecasters to a lower grade on the government pay scale, which resulted in a mass exodus/retirement of dissenting personnel. Less than a decade later, the advent of the Digital Forecast Database and the new forecasting routine not only solidifies forecast offices as collaborative environments but also seeks to extend meteorological deliberation past the boundaries of a single office to encompass the entire network of NWS offices.

Once again, Fine's ethnography serves as the "before" comparison point. Fine notes that interoffice collaboration was largely absent during his fieldwork and that "offices tended to be islands unto themselves" (Fine 2007, 121n82), effectively creating "an archipelago of 122 islands of weather with no overlap" (Fine 2007, 143). That its field office structure was a network in name only arguably constituted a moot point for the NWS administration before the implementation of the National Digital Forecast Database. Until 2003, forecast discrepancies between offices were not likely to be noticeable without a close side-by-side comparison of text forecasts for abutting counties. Now, however, all one has to do is take a look at a Database grid, and the color incongruities make any forecast discrepancies between offices visually jump out. For the Digital Forecast Database to be a success, then, consistency among office forecasts becomes a must. Indeed, one of the milestones set by the NWS administration, to be met before the Digital Forecast Database could officially be released to the public, was meteorological consistency along office boundaries for all forecast grids 90 percent of the time. Once meteorological consistency became operationally imperative, the promotion and regulation of interoffice collaboration turned into an organizational priority. In August 2003, the NWS corporate board approved a "Weather Forecast Office Operations Vision and Philosophy," in which collaboration was featured as one of the three principles of field office operations (NWS 2003a). Already in July 2003, a live NWS teletraining presentation, titled "Collaboration in the IFPS Era" (Howerton 2003), had expounded on the concept of collaboration, which was distinguished from coordination, presumably the process followed by NWS forecasters thus far. Forecasters were told collaboration means "exchanging ideas to meet a common forecast solution," whereas coordination only involves "notifying surrounding offices of your forecast." And they were warned: "while you may be able to 'beat' the collective on a given shift, remember that our ability to reach meteorological consistency in the grids goes to our credibility as an agency."

Today, over a decade later, it is fair to say that interoffice collaboration remains an uphill battle at the NWS. It certainly remains an uphill battle at Neborough, an office that, in the words of a high-ranking Eastern Region Headquarters official, "must come to terms with the fact that two out of the three offices you are collaborating with are dinosaurs." Yet there is no doubt that the spirit of collaboration has slowly started to take root. There is now clear evidence of what Weick and Roberts (1993) call "heedful interrelating": forecasters maintain conscious awareness of how

CHAPTER TWO

their forecast grids fit and contribute to the overall National Digital Forecast Database, and their exchanges with neighboring offices are precisely geared toward accomplishing what is now more readily recognized as a common organizational goal. This "collective mindfulness" (Weick, Sutcliffe, and Obstfeld 1999) has been long in coming, but it is finally bearing fruit because the NWS administration took great care to operationally embed its call to meteorological consistency both conceptually and technologically: workstations were outfitted with a chat room application to allow forecasters to deliberate quickly, simultaneously, and often; consistency became operationalized, and a set of collaboration trigger values were established; the IFPS software suite was upgraded to include an Intersite Coordination Database so that forecasters could upload their grids while still editing them and preview how their forecast draft looked next to those submitted by surrounding offices; and meteorological consensus was, time and time again, cast as the most reliable measure and long-term strategy for ensuring forecast quality. When I left the field in August 2004, getting other offices to collaborate with them was a source of daily frustration for Neborough forecasters.[14] When I returned in January 2008, the improvement in the deliberations between Neborough and its neighbors was undeniable—communication was much more dependable and frequent, and it now consistently included meteorologically based decision making, not mere "wheeling and dealing." To be sure, complaints are still voiced that the push toward interoffice consensus ends up "compromising the science" and therefore the accuracy of one's forecast.[15] Yet increasingly, thanks to the organizational commitment to a seamless-looking forecast, the NWS field office structure actually thinks and acts less like a scattered archipelago and more like a moderately dense network.

These days, as part of her preliminary assessment of available guidance, our forecaster will quickly scan the chatter in the chat room, preset to filter out everything but messages pertaining to the Eastern Region. And once she has formed a preliminary idea of what she would like to do with the forecast, she will share what she considers to be the weather problem of the day and, therefore, her proposed plan of attack, sometimes explicitly requesting feedback from one or more surrounding offices. By and large, forecasters in adjoining offices share the same data on their computer screens. Especially when working the same workstation, they may therefore often prove to be crucial sources of information and decision-making support. To be sure, not all Neborough forecasters heed the official advice to collaborate "early and often" (Howerton and White 2003). A few, in fact, only enter the chat room if they run into a consistency problem. Yet, despite variation in collaboration practices, there is a clear pattern in the ebb and flow of messages flooding the chat room during each shift. The first, and shortest, wave occurs around the beginning of the shift and is driven by announcements posted by a variety of National Centers of Environmental Prediction. The second, and longest, wave occurs about two hours into the shift, when forecasters have a preliminary sense of what they would like to do with the forecast and are ready to talk meteorology with their peers in adjoining offices. Consistent with the newfound spirit of interoffice collaboration permeating forecasting life at the NWS, the greater the uncertainty about the available observational and model guidance for the region, the heavier the chat room traffic. The bulk of the chatter typically revolves around the details of the weather situation in the short term, but an expectation of hazardous weather beyond the first thirty-six hours of the forecast means it will be the long-term forecasters who will colonize the chat room during the shift. The conversation throughout this wave of deliberation is not only dense but also exceptionally cohesive. This is truly a conversation about meteorology and, unlike the early days of the chat room at the NWS, most every forecaster is bound to chime in with a carefully crafted opinion about model performance, atmospheric dynamics, and forecasting challenges. This second wave of deliberations can continue for well over an hour, with the release of the new NAM and the GFS model runs right around this time adding more fodder to the discussion. The third and final wave of chatter occurs toward the end of the shift, once forecasters have begun producing forecast grids and uploading them on the Intersite Coordination Database. Any grid differences between two offices exceeding the prescribed consistency thresholds are automatically flagged by the NWS central server and frequently result in quick and isolated bouts of negotiations that may or may not reach a resolution.[16] In contrast to the previous wave of deliberations, chatter explicitly referencing interoffice consistency involves considerably less discussion of meteorology, and it also reads decidedly less like a proper collaboration and more like a bargaining transaction. At this stage of the shift, the overriding goal becomes to quickly and painlessly fix any remaining issues without overhauling the entire tenor of one's forecast.

Overall, exchanges in the chat room are brief and to the point, just like in-house deliberations. If the conversation grows elaborate between two forecasters, they will usually retreat to a private chat room or pick up the phone. The tone always remains professional and collegial: friendly banter is restricted to social pleasantries,[17] while disagreements never continue past two or three rounds of negotiation. No need to risk violating norms of collegiality or breach chat room decorum when there is a technological work-around to the deliberation impasse. This work-around is known among NWS forecasters as "multismoothing," and it involves exploiting the "smoothing" function of the IFPS—intended to give evenness and polish to temperature, pressure, or wind speed contour plots—to touch up just the edges of one's forecast grids so that they seamlessly blend with the grid edges of the neighboring office. The multismoothing tactic is rarely resorted to openly anymore, a far cry from the early days of the IFPS implementation, but interoffice collaboration dilemmas abound. And there are still times when one must find a way to remain compliant with interforecast consistency policies without sacrificing the quality of one's own forecast.

Doing the Forecast

Our forecaster is now ready to actually start doing the forecast. To be sure, she has been in forecasting mode from the moment she stepped foot on the operations deck. Being briefed by the outgoing forecaster, studying observational and model guidance, deliberating with colleagues—these are all organic parts of her forecasting process. But it is when she is actually faced with the minutiae of the task at hand, when she actually must make decisions for over sixty weather elements, many of them at an hourly 2.5-square-kilometer resolution, that forecasting truly begins. Indeed, there is usually a clear transition at this point of the forecast shift, a switching of cognitive gears of sorts, marked by getting up to stretch one's legs, running out to get a cup of coffee, or fixing a snack to eat.

If one agrees with the pragmatist premise that decision making is grounded in practical action, then surely meteorological decision making at the NWS today is properly constituted through the very act of "doing grids." I do not wish to make too much of this point because forecasters are never entirely disconnected from gridwork by virtue of having to keep the forecast in line with current weather conditions throughout the shift.[18] Yet the difference in the way they qualify their reasoning in chat room discussions before and after they start doing the grids is striking. Chat room statements at the beginning of the shift are thus written in the conditional mood of "would" and "could," and they are typically prefaced with "First guess for tomorrow evening is . . ." or "Prelim thoughts are . . ." One or two hours later, at the end of deliberations but before grid editing, statements still indicate a conditional mood and feature phrases such as "Am leaning toward . . ." or "Current thinking is that . . ." Once gridwork is underway, however, the conditional tense disappears from chat room deliberations and, even though deep in the process of negotiating away discrepancies with neighboring offices, forecasters sound confident, or at least resolute:

{2004/6/22 7:17} Neborough_long: NeboroughNeighbor_long, is it too late to ask you down 2 degrees Fahrenheit for Monday's max temperature? Pretty chilly 850-hPa level trough drifting overhead. Thanks for checking. I think we will be consistent as we are but I'm under you. Don't think GFS MEX MOS can free itself from climatology that far out. 850-hPA air temperature anomalies are minus 2 standard deviations for Monday. A deck of clouds could mess this up but if the ensembles are halfway decent, despite the short night, it will be chilly in some of these interior valleys a week from this morning. Thanks.[19]

Note that I by no means wish to suggest that it is only exceptionally now in the IFPs era, with the tethering of NWS forecasting to gridwork, that meteorological decision making arises in the doing of the forecast. On the contrary, I maintain that weather predictions, like all decisions, represent eminently practical solutions to emergent, locally defined problems and as such have always been highly sensitive to the material conditions of their production. That said, however, it is also fair to say that the extreme physicality of the new NWS forecasting routine—namely, the extraordinary amount of *manual* labor that must now be put into formulating the NWS forecast—throws the embodied, contextual, and pragmatic character of meteorological decision making into the starkest relief.

Granted, in theory, the NWS forecast could now be produced in a relatively swift and effortless manner. The beauty of the Digital Forecast Database is exactly that it is a database: it already stores a forecast by the Neborough office, a forecast that could potentially be left as is when 4:00 p.m. rolls around and our forecaster must submit her forecast. Indeed, the only official dissemination requirement is submitting grids for the new day 7 of the forecast by 12:45 p.m. so that they can be properly mosaicked into the existing database by the 4:00 p.m. deadline, in time for the evening drive-time radio and television news. Otherwise, however, the National Digital Forecast Database is supposed to be weather event driven, not schedule driven. And forecasters have been explicitly advised to "begin [the] forecast process with the idea that the existing collaborated forecast is the best forecast" (NWS 2004b).

Aside from the obligatory appendage of a new day 7 to the forecast, then, the only valid justification for changing the existing forecast grids is changes in the weather. If continuity across offices constitutes a recent phenomenon at the NWS, continuity across shifts is a long-standing staple of forecasting etiquette and practice. Drastic changes in the weather forecast from one shift to the next are easy to spot—and called into question—by forecast users. By extension, "flip-flopping" or "yo-yo forecasting" back to the original forecast carries an even greater burden of proof. In the absence of changes in the weather, forecasting inertia not only helps preserve the credibility of the Digital Forecast Database but it also helps preserve harmony among the forecasting staff. Fine (2007, 138ff.) sensitively documents forecasters' ever-present dilemma to change or not to change the forecast they inherited from the previous shift as they go about putting together their own forecast. This inheritance burden has been considerably eased in recent years because forecasters are now instructed to initialize grid editing with fresh model data instead of preexisting grids so as to ensure higher verification scores (Watling 2004). There is little doubt, however, that social norms of collegiality can be reliably counted on to reinforce institutional pressures toward upholding a consistent definition of the situation.

And yet, even on a quiet weather day like today, when she fully intends to maintain the status quo, our forecaster will still end up toiling long and hard over the grids. Despite the generally temperate and conservative approach to meteorological prediction fostered by the organizational impetus toward one single NWS message, doing the forecast is still a lengthy, laborious, and elaborate process. The reason, simply put, is that weather forecasting at the NWS does not only involve staying one step ahead of the weather but staying one step ahead of the weather models as well. To credibly secure their professional status and autonomy within the NWS, forecasters must constantly prove that computer-generated forecasts are just guidance—that is to say, they must constantly improve on them.

The driving force behind this competition of forecasters against forecast models is the NWS itself. The official metric for tracking the accuracy of the final "human" forecast rests explicitly on a comparison with model performance. More specifically, the verification score of model-generated forecasts has been set as the standard against which the skill of the human forecast(er) is measured (cf. NWS 2011a). Quite literally, then, providing a good forecast in the eyes of the NWS administration presumes "beating the models." And striving to produce the most accurate NWS forecast possible presupposes that one has accepted the organizational challenge to race against the machine:

Forecaster: But the biggest thing, though, is that we just have to beat the models.
Phaedra: Meaning? As far as the NWS is concerned?
Forecaster: Exactly. Well, the Weather Service in general, our office in particular.
Phaedra: That's like the philosophy here?

Forecaster: Pretty much, yeah. Wayne tries to drill in our heads that we have to beat the models or we're gonna lose our jobs. So, gee, I learn: If I can't beat the models, I'm not doing my job! [*laughing*]

In effect, on a quiet day like today, forecasting the weather is decidedly less of a match against nature and more of a match against machine, a numbers game. Of course, to be contenders in this race against the machine, forecasters must work together with computer models toward an accurate forecast. It would be folly to do otherwise, and days where, because of maintenance or a technical glitch, the latest model runs are not promptly available on their screens, forecasters' dependence on computer-generated forecasts is deeply palpable. The trick, then, is to figure out how to harvest and exploit the information contained in models but still make the forecast one's own. That is much easier said than done, but expected nonetheless. Although infrequent, occasions where NWS forecasters were wrong but the models were right are therefore the hardest to live down:

They underdid the forecast for today. When I left Metrocity at 5:30 a.m., it looked like the skies were going to clear out. But by the time I was five miles from the office it was raining hard. When I arrived, Tom (short-term desk, midnight shift) was busy updating the grids. A few minutes later, Peter (short-term desk, morning shift) walked in.

Peter: I guess the GFS was right. I thought it messed up with the QPF [quantitative precipitation forecast] out in [southern state], but I guess it was right.
Tom [*morosely*]: Yup, it was right. The wave came out of nowhere. Things looked dry a couple of hours ago. We went pretty much with the forecast you had yesterday.
Peter: Really?! Wow! . . . My wife was busting my chops this morning. She asked me if it was going to rain and if she should pick up the kids' toys from the deck and I told her no. "You said it was not going to rain but it is raining!" . . . The rain woke me up at 4:00 this morning and I said: "Fuck!" . . . We really got burned by this rain today. The GFS was right.
Arnold (HMT): Well, you win some, you lose some.
Tom: Or something like that.

Fine (2007, 119) reports that, during his fieldwork, veteran forecasters at the Chicago office said they avoided examining model guidance too closely because models were their "worst enemies"; they constrained them. This is no longer possible in the IFPS era. While the NWS has always sought to capitalize on the competition between forecasters and models, the new forecasting routine is expressly designed with that purpose in mind. Forecasters are now physically tethered to models. Recall that the first step of doing the forecast involves importing a fresh batch of model graphics into the Graphical Forecast Editor to serve as the baseline for all subsequent meteorological decisions about a particular grid/weather element. This baseline may consist of an unadulterated version of a single model, or it may be a blend of several models—the matter of which model(s) to start with is left to our forecaster's expert discretion. That NWS forecasters will directly base their forecast on a model, however, has now become a given, scripted into the very technology necessary for producing the forecast (cf. Akrich 1992).

Yet every technological script begets its work-arounds (Gasser 1986; Pollock 2005). Such work-arounds do not have to be radical or subversive, like the practice of carpet bombing, but artfully finessed tactics—and all the more pervasive for that. One need only witness the casual but intense ritual that occurs once grid editing is all but over. Our forecaster prints out the MOS (Model Output Statistics) bulletin of her chosen baseline model and carefully highlights the MOS data for the six cities and towns that serve as official forecast verification sites for the Neborough office. Always referring to the MOS bulletin, she then proceeds to "touch up" the corresponding grid points a couple of degrees/knots/percentage points higher or lower, as the case may be. The reasons offered to me by Neborough forecasters to justify this practice have substantive merit: urban areas generate more heat than surrounding land, higher elevation areas are colder and wetter, coastal areas are windier. Very rarely, however, would forecasters reserve the same special treatment for the remaining forty locations on the Neborough MOS bulletin, some of which are also urban, coastal, or at a higher elevation. The forecast is good enough as is. With a few minutes remaining on the clock, this is the time to go head-to-head with the favored model.

The physical tethering of NWS forecasters to models naturally resonates with their deepest-seated status anxieties. Still, forecasters are apt to only exceptionally be aware of the politics of standardization. Under normal conditions, professional standards are looked on as allies, not foes, in establishing one's credibility and authority (cf. Bowker and Star 1999; Bush 2011). Sociologists, however, will immediately recognize this latest technological scripting of the NWS forecast routine as just one, albeit striking, instance of a much broader and all-encompassing organizational process: the cultivation of meteorological decision making in accordance with the institutional logic of the NWS, as articulated in a set of "material practices and symbolic constructions" (Friedland and Alford 1991, 248; see also DiMaggio 1997; Thornton et al. 2012). Indeed, as we have seen, the new routine has already become normalized among NWS forecasters, its logic of forecasting action having made alternative behavior not merely deviant but nearly inconceivable. But does this mean that the new routine has become routinized? Are NWS forecasters, in other words, so thoroughly conditioned to reading weather information in terms of already docile and tractable problems, so inured to the inherent uncertainty of predicting the weather that their decision-making process has become mostly rote and automatic?

Cultural sociologists insist that culture both constrains and enables action; that it delimits as well as offers up opportunities for resourcefulness. And organizational sociologists highlight how contradictions and inconsistencies in the logics of institutional fields allow for change as well as continuity. When it comes to most conceptualizations of *habitual* action or practice, however, actors appear no different than "cultural dopes," passively carrying out sedimented behavior patterns. It is only when one uses a pragmatist lens to look at it that practice does not magically deteriorate into mindless action but instead retains the recognizable "deliberative attitude" of agency—namely, the ability to creatively reconfigure past organized responses in anticipation of an evolving situation (Mead 1932, 76; see also Emirbayer and Mische 1998; Mische 2009).

So, too, with our forecaster as she proceeds with her forecast routine on this fair-weather day. To find her bearings, she will of course rely as much as possible on the well-tried techniques and other such material and symbolic elements of her stock of meteorological knowledge, accumulated over the years and organized according to the day-to-day requirements of forecasting for the NWS. Yet the more experienced a forecaster she is, the more she is bound to be conscious of the atmosphere brewing with possibilities. The more she is bound, therefore, to avoid a generic, if rule-bound, approach in favor of decision making that is sensitive to the emergent specifics of the situation at hand (cf. Dreyfus and Dreyfus 2008). I will have much to say in the following chapters about the mechanisms that underlie forecasters' improvisational use of their meteorological repertoire. For now, I just would like to underscore the point that weather forecasting practice at the NWS entails adhering to institutionally endorsed decision-making habits but still remaining tactically alert to ongoing weather developments. This is certainly the case at the Neborough office, as we shall see. But it is also what is organizationally expected from all NWS offices. Official guidelines acknowledge time and time again the indeterminacy of atmospheric phenomena and the local contingencies of forecasting action, and they explicitly entrust on the practical judgment of individual forecasters the day-to-day negotiation of procedural standards.

By 4:00 p.m., grid editing should be complete. This long-established dissemination deadline, meant to be in sync with the evening rush hour news cycle, is adhered to strictly by Neborough forecasters even though the NWS forecast has not been directly broadcast by the media for over a decade. It is truly a testament to their conviction that above all else they are public scientists that, during the early chaotic days of the IFPS, they would scramble to send a partly finished forecast to the central server because "people are waiting." These days, the forecast is fully finished by the dissemination deadline, but this still means that after lunch, usually eaten at her desk, our forecaster will enter "crunch time" and, just as in the beginning of the shift, all that will be heard on the operations deck is the fierce clicking of the mouse.

But there is more to be done still. With the forecast grids on their way to the National Digital Forecast Database, there is now the matter of the text forecast to tend to. Sending the *word* out remains an important part of the NWS forecast. Yet it no longer constitutes an important part of the NWS forecasting task. Recall that the legacy NWS text forecast is now automatically generated out of the grids into a more specific but awkward version of its former self: "Monday: A chance of showers before 11am, then a chance of showers and thunderstorms between 11am and 4pm, then showers likely and possibly a thunderstorm after 4pm. Mostly cloudy, with a high near 54. Calm wind becoming west around 8 mph in the afternoon. Chance of precipitation is 50%. New precipitation amounts of less than a tenth of an inch possible."

While waiting for the text generator, our forecaster turns to the final forecasting task of the shift: the Area Forecast Discussion, where NWS forecasters are expected to elaborate on the meteorological challenges and reasoning behind their forecasting decisions. Traditionally a coordination tool among forecast offices, the Area Forecast Discussion has over the years become one of the most popular and widely read NWS products (NRC 2006, 69). Forecast users, from broadcast meteorologists all the way to windsurfers, maintain that by consulting the Area Forecast Discussion they are able to derive a confidence level on NWS predictions. Amid increasing pressures to more effectively communicate forecast uncertainty, the NWS thus made it its policy in 2003 that forecasters shall write their Discussion in plain language—free from the usual jargon, acronyms, and telegraphic sentences—so that it is more accessible to the general public. Despite the extra workload it represents, the Area Forecast Discussion currently absorbs much of forecasters' energy. To it they devote all the care and attention they are not allowed to give the text forecast anymore. Some keep a draft file of their Discussion open at all times, typing in bits and pieces as they are doing the forecast; others routinely issue multiple Discussion updates during a shift; and all make sure to sign their last name at the bottom of the text they so fastidiously composed. In effect, on a quiet day like today, it is through the Area Forecast Discussion that forecasters feel they are sending *their* word out. Knowing they have a captive audience, meanwhile, fuels the gusto with which they embrace the task. Consequently, Area Forecast Discussions come to acquire distinct personalities, to the point where one need not see the name to know which forecaster authored which Discussion. And they predictably acquire a following, with users reporting they have a higher confidence in the forecast when X or Y is working the shift. More importantly for the purposes of this analysis, the act of composing an insightful synopsis about the weather problem of the day and expounding on their decision-making rationale may sometimes compel forecasters to rethink one or more of their forecasting decisions. It would thus not be out of the ordinary to find them going into the grids for one last time after they have finished articulating their Area Forecast Discussion.

It is rarely before 5:00 p.m. that our forecaster hits the "Enter" button for the last time. Now she gets up, stretches her back, and starts walking toward the cubicles in the back of the operations deck, calling out to the incoming forecaster that she is ready to do the briefing. Together, they will walk back to the screens, the outgoing forecaster will take her seat, and the briefing will begin.

THREE

Distilling Complexity: Atmospheric Indeterminacy and the Culture of Disciplined Improvisation

The Neborough region is famous for its exciting and changeable weather year round. The winters here are long and fierce, with legendary blizzards and snowstorms; the springs are prone to flooding events; and the summers are hot and humid, with frequent thunderstorms and the occasional tornado or even a hurricane. The coastline not only complicates the picture by adding the ocean effect into the mix but it also comes with its own set of weather hazards. While I never got to experience extreme weather during my twenty-two months at the Neborough office, weather I saw abundant. And I saw it in ways I had never seen weather before.

This chapter delves into the complex ways NWS forecasters take stock of the weather. Sharpened by the indeterminacy of the atmosphere, the need for situational awareness cultivates in forecasters an omnivorous appetite for information—even compelling them to momentarily step away from their workstations and enlist the weather outside in order to prevail over the ambiguity of the weather on their screens. But how are they to then harness this disparate information and project themselves into the future? How, in other words, are they to distill and extrapolate complexity into a provisionally coherent weather forecast?

Fifty Ways to Ground Truth

One of the running jokes about the NWS, relayed to me by a commercial fisher in the region, is that the only windows in NWS offices are Windows 95 (apparently this is an old running joke). True enough, NWS forecasters do spend most, nearly all, even, of their time in front of computer screens, busy looking at and manipulating weather graphics. After all, it is only on their computer screens that they can see along the vertical column of the atmosphere or beyond their immediate surroundings, let alone behold larger-scale meteorological features such as fronts or jet streams. It is only on their computer screens that they can evaluate how model-generated forecasts are "catching up with reality" by visually superimposing model and observation data. The laboratory-like environment of their office has indeed become their natural environment. And for their part, NWS forecasters are certainly apt to dwell on the high-tech dimension of the forecasting task, especially in the presence of outsiders, presumably to drive home the complexity, but also the scientific legitimacy, of the undertaking.

Yet that would be a rather facile account of the weather forecasting process. The streamlined appearance of the operations deck belies the fact that forecasters routinely contend with diverse, partial, nonoverlapping, and often contradictory pieces of information. As preoccupied as they are with the weather pulsating on their screens, they still remain always attuned to a wide variety of inputs and stimuli about the weather. Consider the following excerpt from my field notes for March 12, 2004.

8:00 a.m. Simon has taken over the short-term desk for a while so that Peter can tie up some administrative loose ends. He issues a couple of nowcasts for snow squalls and rain or snow showers. An hour later, looking at the observation reports on his screens, he exclaims: "34 sustained winds at X-town for 10 minutes!" Subsequently, he amends the previous nowcasts, adding a line about wind gust potential of 35 to 40 m/h.
. . .
9:30 a.m. Spotter reports are coming in regarding wind gusts. One meteorologist at Channel 12 in Y-town just called to report gusts of 47 m/h.

Peter (back on short-term): Consistent with the observations.
Simon: Sorry I left you in a lurch, Peter. I thought I had things under control. 38 would
    be as high as they would go. Maybe I should have issued a wind advisory.
Peter: That's alright. I do think this is marginal.
. . .

CHAPTER THREE

1:00 p.m. A sound is heard on the roof. Peter: "That's weird." He checks the radar and the satellite imagery and then gets up and walks toward the office kitchen, trying to catch a glimpse of the weather through the windows on his way there. He opens the glass door next to the kitchen and stands there looking outside for a good twenty seconds. Returns to the operations deck.

Simon: What's going on?
Peter: Chilly rain.
Simon: I'm sorry, Peter. I thought with gusts to 48 that's all we were going to get. Little did I know . . .

No more than five minutes later, Cody walks in and announces "It's snowing!" Simon is apologizing profusely.
. . .
1:30 p.m. Cody is standing outside Simon's office, which has a rather large unobstructed window view, looking on operations while sipping his soda. From there, he cries out peering into Simon's office, "What the hell is that?!" He goes outside through the door near the kitchen and is trying to figure out what kind of precipitation it is. After a few moments, Simon gets up: "Okay, I'll bite. Let's see what this is." I follow him outside, where he and Cody pick up some snow from the ground and examine it, rolling it around in their hand and crushing it with their fingers. They tell me it's snow pellets—they look like tiny opaque grains of ice. The two men get back inside. Simon brings a snow pellet back to Peter. He is frustrated. Keeps apologizing, "I goofed."

This episode is punctuated by a host of diverse weather observations: wind measurements recorded by NWS automated weather stations; wind measurements recorded by unofficial weather stations and reported over the phone by citizen volunteers and a TV meteorologist; radar images; satellite images; a raining sound on the roof; passing glimpses through windows; a personal 180-degree visual and sensory assessment of the surrounding conditions; and close examination of frozen precipitation specimens. This intermingling of weather observations is considered usual practice at the Neborough office. Indeed, it would appear to be usual practice across all NWS field offices, including the Chicago office (Fine 2007, 28–31). For, even under routine, benign weather regimes, when forecasters can rely on constant up-to-date feedback from NWS weather stations and buoy sites, satellite, and radar—that is, information processed, packaged, and transmitted directly to their screens—they are still faced with limited measurements made by limited sensors. The threat of hazardous weather only serves to increase the number and variety of information sources brought to bear on the situation, thus exponentially compounding the challenge of the undertaking.

I will return to the management of different meteorological scenarios in a moment. For now, I just want to point out that the sparseness of weather data sites, the imperfections of weather instruments, the high dependency of weather prediction models on initialization conditions due to limited understanding of the atmosphere, all these factors compel forecasters to go looking for weather information beyond their computer screens, wherever they can find it. Indeed, one could accurately describe them as weather observation *omnivores*: they go to considerable lengths to procure an assortment of weather information, every morsel ostensibly taken into account and weighed against all other available information before it is ultimately included in or discarded from the pool. Tellingly, not only are all weather reports indiscriminately lumped together under "weather obs," but they are all considered part of "ground truth." A military term originally used to refer to data collected on the ground, where the action is, as opposed to data from remote sensing equipment, ground truth in the context of weather forecasting operations is vague to the point of vacuousness, as reflected in the following definition by a Neborough forecaster: "Ground truth would be an observation by instrument or by someone who knows what they're looking at."

To be sure, forecasters' taste for weather information, albeit eclectic, is quite discriminating. This they like to attribute to meteorological "common sense." Yet the naturalization of this knack at sense making in the face of equivocal atmospheric information represents an enormous organizational feat (cf. Weick 1995), the result of the institutionalization of particular logics of seeing and telling the weather that make alternative solutions all but inconceivable in practice. Despite their claims to the contrary, Neborough forecasters are drawn to particular sources and types of information not on the basis of some sort of objective assessment of the weather pattern at hand but rather because they have been primed to recognize particular weather patterns as such in the first place. When I asked them to rank observation sources for a number of weather scenarios, they all predictably employed the same ordering criteria and consistently ended up with the same hierarchies, despite the multitude of sources and measurement biases to consider. If some found it difficult, even frustrating, to come up with a ranking order, it was only because they were forced to consciously reflect on and articulate how exactly data figured in their decision-making process. Only the newly arrived Neborough intern was unable to complete the exercise, his masters in meteorology notwithstanding.

CHAPTER THREE

Sorting through the welter of weather reports is thus part and parcel of being a weather forecaster, a know-how developed and continuously honed as one gains experience in the field. Becoming "intuitively" drawn to and developing a taste for particular data and problem solutions hinges precisely on having become socialized into a particular decision-making regimen designed to foreground particular habits of human judgment as epistemically virtuous, aesthetically agreeable, and pragmatically prudent. At the NWS, the logic of this regimen is partially externalized in the so-called NWS Directives: over a hundred policy and procedure regulations regarding forecasting the NWS way, stored in binders for easy reference at every forecast office and a regular feature of the teletraining courses forecasters are required to complete every few months.

It is thus very telling about the culture of forecasting at the NWS that the intermingling of weather observations during the forecasting task is not simply usual practice—it is in fact institutionalized practice, recognized and promoted by the NWS Directives themselves under a specially coined name: the Total Observation Concept. One encounters a brief explanation of the term in reference to the production of airport forecasts:

> The aviation forecaster must have certain information for the preparation and scheduled issuance of each individual [airport forecast]. All weather elements need not be provided completely and/or at all times in the hourly/special observation itself. Forecasters will also make use of supplementary, complementary and/or augmented observation data, as well as other observing systems (satellite, WSR-88D radar, profiler data) in preparing and monitoring [airport forecasts]. This approach, to issue and maintain [airport forecasts] using multiple, integrated data sets in addition to hourly and special observations, is known as the Total Observation Concept. (NWS 2005a)

The introduction of the concept of the "total observation" within the context of airport forecasts is hardly coincidental. Recall that NWS automated weather stations are located in airports, having replaced human observers in the late nineties. The controversy surrounding the automation of observations is long dead given the prohibitive cost of alternate solutions, even if the issues that spawned it are still alive today.[1] Forced to choose, forecasters say they prefer weather instruments over human observers, even though the former can only measure what is directly above them, because they are reliably available around the clock and can be counted on to always have the same biases and limitations. Whenever given the option, however, forecasters will consult both machine-derived and human-derived observations. The four major airports in the Neborough region, for example, have reports from the automated weather station on site "augmented" by privately contracted human observers, while at airports with no dedicated personnel, traffic controllers are expected to step in as weather observers during rapidly changing conditions. The line between machine and human weather sensors is becoming increasingly blurred in NWS forecasting operations.[2] But the underlying challenge, formally encapsulated in the NWS definition of the Total Observation Concept, persists: on the one hand, there is a need for a holistic, albeit never complete, capturing of meteorological conditions; on the other, the available data represent measurements of discrete variables at specific locations derived from at best complementary sources that often need to be manually supplemented and augmented with additional data.

Despite the scientific tone of the NWS definition, the main message comes through loud and clear. A weather report can at best provide an elliptical, fractured glimpse of meteorological reality, so that forecasters are expected to turn to other sources when they deem they do not have enough information to pronounce the weather. This is not the description of a work-around but of "business as usual." That certainly is how the official vernacular becomes elaborated in the field. Says a Neborough forecaster,

Just because you got data, it does not necessarily mean you know exactly the state of the atmosphere at that time, because you still just have the discrete data points. We take what we call the Total Observation, the Total Observation Concept, where we try to integrate data from different sources. We will in a subjective way rate it to a degree depending upon our knowledge of where it's coming from. Example: somebody just calls in a wind speed report from their Davis anemometer on the roof. We are still going to factor in this information, it's still going to tell us something, but we will kind of have in our minds a much broader error bar on it, and we are going to be a little more suspicious of the quality of that data than say from [NWS automated weather stations], which in terms of ranking will generally be pretty high up on the scale, unless it is precipitation. And we know precipitation is not something that [the automated weather station network] does well, especially during the winter time. But we know it does a pretty good job when it comes to measuring wind, and temperature, dew point, pressure—those parameters it does very well. And it's equipment that's regularly checked and calibrated, so. . . . We will take ground truth from a variety of sources, but we will weigh it to some extent on its quality.

NWS efforts to cultivate a particular meteorological aesthesis among its forecasters as a way of coping with the uncertainty of the subjunctive mood have thus culminated in an epistemic culture of what may be best described as *disciplined improvisation*—an epistemic culture that does not simply intuitively engage in but actually celebrates bricolage (French for "tinkering"), to the point of proactively and methodically improvising in the collection and use of resources during the production of the weather forecast.[3] In fact, the NWS Directives make a point of stating explicitly that forecasters are protected from legal liability when employing the Total Observation Concept, under the discretionary function exemption of the Federal Tort Claims Act (NWS 2009b, 6; see also Klein and Pielke 2002).

Still, beyond lending institutional legitimacy to forecasters' omnivorous practices and creative tinkering by affirming that they "can and should" identify, collect, and use disparate data (NWS 2009a), the NWS Directives have little to say about how exactly forecasters should go about distilling all this complex and contradictory meteorological information into a coherent prediction. How disciplined improvisation is to be transformed into a masterful weather forecast appears to fall under forecasters' sole responsibility and discretion. Yet, as Fine (2007, 177) aptly points out, sorting one's way through this heterogeneous mass of ground truth reports amounts to little short of "an act of professional heroism" because forecasters typically find themselves "hostage" to truth claims well outside their direct purview. To this invisible "articulation work" (Gerson and Star 1986; Clarke and Fujimura 1992) of aligning organizational requirements with the day-to-day coordination of meteorological decision making I will turn in a moment. But first there is another matter to consider about which the NWS Directives are similarly silent. Namely, the role of NWS forecasters as ad hoc weather observers.[4]

Being There

Weather forecasters operate in the "swampy lowlands" (Schön 1983, 42) of meteorology. They have to muddle through critically important meteorological situations that typically defy a technically rigorous, neat solution. This is a challenge facing all professional practitioners, who must confront the messiness of everyday life as a matter of course, but it is especially pronounced in the case of weather forecasters because they are charged with protecting the public from a phenomenon that, while perceptibly present outside, cannot be physically manipulated under controlled conditions inside. Consequently, forecasters are forced to carve a place for themselves between the laboratory and the field—between the preprocessed weather on their computer screens and the weather in the wild—in order to attain a better, fuller grasp of the prevailing meteorological conditions. Phenomenologist Maurice Merleau-Ponty (1962, 302) has famously considered the example of visitors at an art gallery, instinctively repositioning themselves in front of a painting until they hit on the optimal viewing distance, to illustrate the primacy of sensory perception in decision-making action. It is precisely this visceral need to achieve an optimal gestalt, or "maximum grip" (Merleau-Ponty 1962; Dreyfus 1992), on the atmosphere that in practice compels NWS forecasters to oscillate between different ways of viewing the weather. Cast in this light, their habit of leaving their workstations to study the weather outside becomes central to understanding how they impose order out of the disparate and ambiguous fragments of information at their disposal.

As historical studies of science remind us, there is nothing inevitable about the evolution of scientific inquiry into the two distinct idioms of laboratory science and field science (Kohler 2002, 62). Not too long ago, in fact, the nonreferentiality of laboratory science was considered a liability, not evidence of direct access to some universal laws of nature. To secure the necessary cultural and political endorsement and validation, therefore, labs were carefully placed in very particular settings, inhabited by very particular people, such as gentle(wo)men's houses (Shapin 1988), pubs (Secord 1994), monastic workshops (Jackson 1999), or the country estates of the industrial bourgeoisie (Schaffer 1998; Forgan 1994). It is arguably similar considerations that prompted the NWS to embed its forecasters within the communities they serve. NWS forecasters, for their part, have certainly been quite successful in availing themselves of their spatial and structural position as brokers between the organization and its constituencies.[5] The political advantage of developing strong local connections notwithstanding, however, the *epistemic* advantage of personally conducting weather observations to produce a forecast for a thirty thousand square mile region is not immediately obvious. Yet, during my fieldwork at the Neborough office, there were countless, indeed daily, instances where a forecaster would leave his workstation with the express purpose of checking on the weather outside, fully aware that a colleague had just been outside for the very same reason. Being there to see the clouds or experience the wind for oneself was somehow important.

Earlier, Peter (short-term desk) went outside. He opened the door, popped his head outside, and took a long deliberate look around. I asked him why. "For the clouds." Apparently, he was in the process of putting together the airport forecast for Metrocity Airport and wanted to see whether the majority of clouds was stratocumulus or whether there was a fair amount of altocumulus. Arnold and I had just been outside for the [obligatory] 2:00 p.m. observation.[6] Peter didn't necessarily know. I told him.

CHAPTER THREE

He said, "I wasn't with you. I didn't see them." Tom (long-term desk) commented, "This is for the people who think we screw the forecast regularly, the people who say what we're forecasting isn't reality." Dick, playfully: "Reality? What reality?"

* * *

Went outside with Bruce to do the 2:00 p.m. observation. We had just returned to the operations deck when Phil (short-term desk) got up and left his workstation. A couple of minutes later, he came back and announced, "It's really blowing right now!" I said, "You could have asked Bruce, we were just outside." Phil said, "I know, but I like to experience it for myself."

Sometimes, stepping outside can be a way of taking a break from forecasting, of course. As one Neborough forecaster confided, "I go out to cool out, get a sense of the weather. Plus, like when others go out to have a smoke, I want to clear my head, think about girlfriends, look at the sky." But even then, monitoring the weather never ceases. Forecasters are always monitoring the weather; it seems to be a constant undercurrent in their daily lives. This should not be simply attributed to their love of meteorology; rather, it is their past, present, and future forecast that is on the line. And so, whether on a family outing, while at home, on their way to the office, or on their way to the office restroom, forecasters are constantly keeping an eye out on how the weather is "materializing."

To be sure, forecasters' unscripted observation outings are no more important than any of the other weather reports at their disposal. In line with the NWS concept of the total observation, these outings are yet another means of gaining situational awareness. Indeed, when taken alone, they can be just as misleading as any other source of ground truth.

Cheryl was looking at a couple of suspicious cells on the Neborough radar that "could be just [military] chaff" given the beautiful day out. Couldn't decide. She pulled up the neighboring office's radar, and said, "Yup, looks like they are real. It's possible."

As a rule, then, Neborough forecasters will *not* go outside for the express purpose of assessing current weather conditions: if they are not working on the forecast for the next thirty-six hours; if they are working on localized and short-lived weather events not in the vicinity of the office; if the weather information on their computer screens appears clear-cut, makes sense, "hangs together." However, when, while in the process of issuing the short-term forecast, they encounter what they deem to be a complex weather scenario or they receive an unexpected weather report, they are bound to instantly glance out the window from where they are sitting, then double-check radar, satellite, and observations on their computer screens, and, *if warranted*, get up, approach one of the two nearest glass door exits, and, as per the earlier episode from my field notes, either pop their head out to take in a 180-degree view of their surroundings or step outside to fully survey the situation.

Meanwhile, sight but also hearing and touch are all instrumental in establishing the identity of the given weather conditions: Peter hears a "weird" sound on the roof and decides to go outside; he sees the rain but he feels the "chilly rain"; Cody and Simon see and feel what they ultimately determine to be snow pellets. In the same vein, maintaining situational awareness does not only involve keeping an eye out for the weather outside; it involves keeping an ear out as well.[7]

John *doing his airport forecasts; stops; to me*: Was that thunder?
Me: I didn't hear anything.
John: [*Waits, listening carefully.*] Nope, just a motorcycle, maybe Jim's.

* * *

Margaret: Is it raining? [*Looks at the radar.*] Yes, could be.
Me, *looking out the window from my seat*: Really?
Margaret: Sounds like it to me.
Me: I'll go take a look.
Margaret: Maybe it's not.
Me: [*I go to the glass door.*] Yeah, you're right, Margaret. Good call! I thought it was the AC.
Margaret, *pointing toward the ceiling*: All these years in this office, you learn to distinguish the sounds.

* * *

It's been raining hard in bouts and we can hear it on the roof. That is, when the precipitation is light we cannot hear it. We started this game with Arnold—every time we hear the sound we both cry out, "It's raining!" Anyway, at some point, Arnold abruptly got up and announced, "This is not just rain, this is wind, too!" True enough, once he said it, I noticed the windy rain whooshing on the roof. Arnold approached the glass door, opened it, and peered into the darkness, trying to make out what was going on.

Expert problem solving and expert cognition more broadly are inherently embodied processes. The mind is thoroughly constituted by bodily activity even in the compulsively artificial and aseptic environment of a laboratory. In fact, the more technologically advanced the decision-making environment, the more "transparent" and natural the technologies in it (Clark 2003) in order to better allow decision makers to pick up the hands-on physical expertise or "muscular Gestalt" (Dreyfus 1992, 249) necessary to smoothly and flexibly find solutions to the tasks at hand.[8] That weather forecasters are wont to employ their body as an additional weather instrument in order to arbitrate the complexity of the atmosphere is therefore but one indicator of embodied—that is to say, expert—forecasting skill. The lab-like environment of NWS operations decks is not only meant to sterilize and enhance the weather, filtering out the noise of the field; it is also expressly designed to provide forecasters with the necessary technological "scaffolding" (Orlikowski 2006; Clark 1998) to extend their meteorological perception so that they will be able to see further and better into the future of the weather.

To understand, then, how forecasters master uncertainty, one must be attentive to how their bodies figure in the forecasting task, processing the weather information streaming inside the office but also the ambient weather floating outside. The possession of skill—that is to say, embodied confidence—in both domains simultaneously is not a given. A veteran forecaster, all Neborough forecasters agree, will need at least a couple of years to become sufficiently attuned to the features of a new meteorological microclimate, while a local intern will need an equal amount of time to become sufficiently conversant with the techniques and tools of the forecasting task. Both types of experience are necessary for developing an authoritative meteorological expertise. In effect, being a good NWS forecaster hinges on the *in-corporation* of multiple and differing logics of knowing the weather.

To be sure, echoing the various epistemic cultures of the laboratory, and unlike the field sciences, weather forecasting does not assign a primary research function to the senses (cf. Knorr Cetina 1999, 95). This is reserved for technology, which is seen as mediating between the body and the phenomenon under study. At best, weather forecasting resembles the epistemic culture of biology, which may put a high premium on manual dexterity and physical stamina but only pays tribute to embodied skills in an undifferentiated and assistive capacity (Knorr Cetina 1999). Still, emulating the culture of the laboratory can only go so far in the case of weather forecasting. While dependence on the computing power and visualization capabilities of machines is necessary to master the physical complexity of the atmosphere, the ultimate objective, always, is to adjudicate on the *social* consequences of the atmosphere. Atmospheric dynamics, after all, are articulated as "the weather" only to the extent that they are considered socially salient. What matters, and what is featured in weather forecasts, is the socially marked subset of atmospheric phenomena, what forecasters call "sensible weather"—the tangibly real weather that we all get to experience. As it is, the first to be skeptical about the correspondence between their predictions on the screen and what is transpiring outside are the forecasters themselves (cf. MacKenzie 1990, 372). When "in limbo," therefore, they find themselves compelled to step outside in search of further clues to resolve the equivocal evidence inside the office.

As one would expect, the threat of hazardous weather makes the need to resolve the indeterminacy of the atmosphere that much more poignant and the urge to look outside in search of further clues that much more frequent. But the persistence of these unscripted outings even under routine forecasting operations cannot be overstated.

Tom, chatting online with a forecaster from a neighboring office: "I decided not to include rain in my [forecast] for 2 reasons: #1 it will wrap up very quickly. #2 it is very light. I went outside to check on this before sending my [forecast], and while I could see very small droplets in the light, I could not feel them."

In this example, Tom sees on his computer screens that the rain "will wrap up quickly" and he gets a good indication that it is going to be "light." Going outside, he confirms that not only is the rain very light ("very small droplets in the light"), but that it is so light he cannot feel it. In this respect, his embodied experience of the weather outside complements and enriches the weather information on his screens. But this episode does not simply illustrate the material complexity of the weather. For Tom makes an important forecasting decision based on his experience of the weather outside: because he cannot feel the rain, he concludes it is not worth forecasting. Because he cannot feel the rain, he decides that others will also not feel the rain and therefore will not see this *as* rain. Consequently, he decides not to see it as rain either. Stepping outside allows him to more fully interject himself into the data and distill them into a projection that, he projects, will be meaningfully relevant to his audience.

What will *count* as weather, then, can be just as challenging a forecasting task as predicting the future physical state of the weather, and it is part of the considerations that, consciously or not, inform a forecaster's decision-making process. I will delve into the implications of this insight in the next chapter, where I discuss hazardous weather forecasting. For now, suffice it to say that any prediction about the weather is also a commentary on its social ramifications. That is why NWS forecasters strive to move beyond the available variable-based information inside their office toward a more holistic, *experiential* appreciation of meteorological conditions. That is why being there with the weather matters.[9] And that is why when their predictions do not materialize, forecasters are wont to attribute the miss to a failure to carry out a total observation, a failure to be with the weather:

> Wayne and Phil are discussing yesterday's "gaffe." Basically what happened was they were "buried in forecasting," and although they had reissued a nowcast for patchy fog, they didn't add a line for black ice. Wayne blames himself. He had been too excited with their "forecasting success" of the previous day.
>
> Phil: Well, what can you do? It's in the past. We have to move on. . . . Another thing is, too, black ice is not really weather.
> Wayne: I know what you mean. It doesn't fall off the radar. But what we should have done yesterday is just looked outside, walked around the parking lot around 5:00 a.m., and then we would have gotten a sense of what was going on.

Anthropologist Tim Ingold has recently argued that the language and concepts we use to account for how we inhabit the open air consistently fail us. Even the shift from a representionalist to a materialist idiom, says Ingold (2007), poorly captures the fact that we do not dwell on a solid surface of people and things but rather *within* a "weather-world" of flux and flow. "Strictly speaking, the weather is not what we have a perception *of*; it is rather what we perceive *in*. [The weather] is not so much an object as a medium of perception" (Ingold 2005, 102; emphasis in the original). Ingold's critique restricts itself to popular and social scientific discourses about the experience of the weather and does not take into consideration the highly sophisticated knowledge infrastructures and cognitive heuristics weather forecasters are working with. Nevertheless, he evocatively sums up the tension inherent in any attempt to reconstruct the outdoors indoors. NWS forecasters are fully aware of this tension—because of their professional training and their daily experience of needing to come up with practical solutions to complex forecasting challenges. Hence they have resolved to be both indoors and outdoors, perceptually emplaced between the weather on their screens and the weather in the wild.

The Total Observation Collage

According to Latour (1999, 28), "scientists master the world, but only if the world comes to them in the form of two-dimensional, superposable, combinable inscriptions"—if it comes to them, in other words, in the form of the spatiotemporally well-nested data that laboratories are set up to generate in the first place. Even amid the rawness of a botanical expedition, Latour's metaphors for the diagnostic process involve scientists "shuffling cards" or calmly arranging "the pieces of [a] jigsaw puzzle" (Latour 1999, 38). The inevitable messiness and tinkering of field science has already been tamed into the orderliness of laboratory science.

True enough, NWS forecasters inhabit a meteorological "center of calculation" (Latour 1987) to which an agglomerate of observation networks continuously delivers a multitude of preprocessed, digestible bits of weather information. And yet it is not by retreating into the depths of their office that forecasters master meteorological uncertainty and establish their credibility. Despite their best efforts to destabilize inside/outside from within the walls of their operations deck, the field resists translation into a specimen, hindering the referential circuit from the weather in the wild to the weather in the laboratory and back again. The dearth of reliable information and the limitations of weather sensors, be they human or machine, led them to develop an appetite for a veritable smorgasbord of data. Far from equivalent, like a deck of cards, or complementary, like jigsaw puzzle pieces interlocking neatly together, the meteorological information forecasters amass in front of them instead includes disparate, overlapping, redundant, and uneven bits of data.

That is why I consider the metaphor of "collage" a more useful heuristic for appreciating how forecasters achieve coherence in the face of uncertainty.[10] Collage—etymologically derived from the French verb *coller*—to glue, to paste together—is associated with the very beginnings of the modern art movement in the early twentieth century. A theory as much as a technique of visual representation, collage challenged the false preciousness and illusionist naturalism of "fine art" by puncturing the uniformity of paint with "real" objects to create a new reality: the image in itself. It allowed artists "to re-enter the world by embracing its artifacts; anti-art became art and art strove to become a part of life" (Janis and Blesh 1967, 11). "The bits of paper I have stuck on my drawings have likewise given me a feeling of certainty," writes Georges Braque in 1917.[11]

For the purposes of this discussion, I define *collage* as the progressive addition of an unlimited variety of information fragments, pictorial and otherwise, of different scale and texture, arranged in such a way as to conjure a new meaningfully whole image, a new gestalt. As a technique, collage is improvisational, open-ended, welcoming of surprising or paradoxical encounters. It accommodates notions of relationality, multiplicity, performativity, reflexivity, material heterogeneity, uncertainty, and emergence. As a finished product, a collage's appeal lies in its poetics of order out of seemingly disparate elements. Collage is at its best when "things relate but don't add up" (Law and Mol 2002, 1).

Conceptualizing weather forecasting as the art of collage is to say, in effect, that weather forecasters distill the complexity of the atmosphere into a coherent account through a process of assembling, appropriating, superimposing, juxtaposing, and blurring disparate pieces of information. As already noted, the organizational mandate for a total observation makes NWS forecasters highly proactive at information bricolage, assembling a wide variety of weather information to supplement what is automatically delivered to their screens via the NWS supercomputer. While messy and unwieldy because not technologically intrinsic to the forecast production process, much of this additional information is nonetheless readily relatable because it is communicated as a report about the weather. Yet forecasters also appropriate downright extraneous information fragments, such as road condition reports faxed over hourly from the Department of Transportation or keyword-filtered news items delivered daily by a press-clipping service. The boundaries of appropriateness are not limitless, improvisation is largely disciplined and habit bound, but the hunt for new "found" information sources is ever present.

This bounty of compelling but disjointed information forecasters seek to organize into coherence through screenwork—though an iterative, laborious process of combining and leveraging the assembled, official and appropriated, materials at hand in novel ways that capitalize on the expansive "real estate" and capabilities of their computer screens. Upper air observational slices are thus superimposed, layer upon layer, in an effort to digitally reconstitute the vertical column of the atmosphere. Hemispheric and local surface weather conditions are brought into one via the zoom in/out function. Atmospheric phenomena are clicked in and out of existence, extended forward in time in a 32-frame loop, and sampled together in a variety of plausible combinations. It is, however, when they turn their computer screens into "surfaces of tense juxtaposition" (cf. Law 2007) that forecasters reap the most benefit. Colligating information that does not at first glance belong together, they are striving to see through the complexity of the atmosphere and distill it into a reasonable representation of the sensible weather. The simultaneous juxtaposition of otherwise separate information bits of differing scale, granularity, perspective, and scope placed side by side on the same or adjacent screens allows them to perceptually group them together into a new gestalt and summon up a unitary reality of a higher order.

But the heuristic value of collage as a metaphor of how weather forecasters craft coherence out of improvisation does not end with the diagnosis of the atmosphere. Recall that thanks to the implementation of the Graphical Forecast Editor in 2003, the mental grouping and distillation of complex meteorological information into a prognosis has become a highly externalized cognitive activity at the NWS. As a result, all manner of screenwork forecasters engage in is directly or indirectly geared toward the production of the NWS forecast. The customary practice of data analysis through juxtaposition should thus not be considered on its own, as some sort of diagnostic countermeasure against confirmation bias. Model prediction graphics may be routinely lined up and pitted against one another as if in a "beauty contest" (Fine 2007, 116), ostensibly to evaluate which model has a better handle on an emerging weather situation. But rarely did I witness a Neborough forecaster go against organizational wisdom—crystallized via intra- and interoffice deliberations—regarding the overall pattern of an impending weather event. Rather, the point all along is to lay the groundwork for the practical task at hand: doing the grids. The juxtaposition of observational and model data primarily serves to provide answers to such seemingly second-order decisions as, "Should I use a blend of the NAM and the Euro to populate the snowfall grid for Tuesday night, or will the NAM do the job?" or "Which model has a better handle on the timing of the rain showers for southwest Neborough Monday afternoon?"

Yet, often unwittingly, in the process of working out a practical solution to these comparatively lesser decision problems, forecasters transform the tenor of the previously agreed-on meteorological pattern in significant ways, simply by introducing nuance and dimensionality into some of its individual features. Because atmospheric dynamics are more or less strongly coupled, the consequences of combining and recombining local information can be striking, cascading beyond the confines of a single weather element or office.

Doug (short-term desk) was deep in his grids when he suddenly sat back in his chair and exclaimed, "Wow! That's too much white!"

Margaret (long-term desk): Let me guess: QPF [quantitative precipitation forecast] trouble.
Doug: I thought the NAM and HPC [Hydrometeorological Prediction Center] had the right idea with a point five six-hourly QPF, but they are trending colder than warranted by latest obs[ervations]. I just ran the Snow Tool, and it's giving me a snow total of nine inches for Metrocity!
Margaret: Oof! Overplayed much?

CHAPTER THREE

Doug: Looks like it's not picking up on the warm air mass very well. . . . Back to the drawing board.

He consults the latest surface observations and satellite imagery, double-checks the NAM temperature forecast and soundings, and then goes back into the Temperature grids, where he uses the Pencil Tool to draw a three-degree higher temperature contour around Metrocity for tomorrow morning and, after some trial and error, ends up increasing the Minimum Temperature grid by two degrees across the Neborough region for tonight. He reruns the Snow Tool and is satisfied with the new snow total output, which, at seven inches for Metrocity, represents a compromise between his original thinking of five inches and the NAM QPF. But "just in case," he tells me, he also runs a "sampler" of the now automatically generated text forecast to make sure his grid changes "amount into one mass of consistency." Finally, turning to the chat room on his middle screen, he informs neighboring offices that, "after fiddling around with the grids," he thought better of it and has decided to include Metrocity in the Neborough winter storm warning for tomorrow.[12] His new forecast in turn prompts the office to the south, "sitting on the fence" up to that point, to drop the winter storm advisory for its northernmost counties.

It is not disciplined improvisation as such, then, but the marshalling of existing resources into a provisionally coherent composition that delineates the space for strategic action. How this process will play out in practice can vary quite dramatically, of course, depending on the perceived risk and scale of the weather situation at hand (see chap. 6). Just as importantly, however, it is defined by the logic of collage, now transmogrified into the "cut and paste" philosophy of contemporary digital culture. "The avant-garde became materialized in a computer," writes new media theorist Lev Manovich (2001, 258, 143): what used to be exceptions became the normal, intended techniques of digital compositing, embedded in the commands and interface metaphors of computer software. By combining different model elements into a single weather grid image, by cutting and pasting graphics from different time periods, by manually redrawing contour plots to incorporate radar, satellite, and surface data, or by introducing new grid point values according to the latest spotter reports, forecasters are able to enlist high and low resolution visualizations, raw and processed data, graphical and textual information, official and unofficial measurements, surface and upper air observations in the production— better yet, the *Gestaltung*—of a weather reality that endeavors to stand up to the indeterminacy of the atmosphere and the weather requirements of their multiple audiences.

This is not to say that the end result preserves the traditional collage aesthetic of juxtaposition and unevenness. Quite on the contrary, the NWS forecast adheres to the exquisitely layered collage aesthetic of the digital age. For the longest time, while forecasters are working on their screens, the forecast betrays the signs of collage activity: jagged edges, discontinuous color patches, uneven pixel sizes. As the forecast submission deadline draws near, however, messiness is transformed into smoothness. At the final moment, the conventional cut-and-paste techniques of assembling, appropriating, superimposing, juxtaposing, and blurring are supplemented by techniques of color remapping and dithering, image morphing, and resolution compensation.

Twenty-first century technology thus helps restore the semblance of nineteenth-century scientism precisely because it relies on collage techniques to do so. Handcrafted-looking weather maps have given way to seamless digital composites, while the rich heterogeneity and discontinuity of meteorological information is conveniently tucked away in a variety of embedded hyperlinks that users are expected to click on to access localized aspects of the forecast.[13] Only television weathercasts seem to still have a use for the old-fashioned collaged look. In the name of infotainment, one still comes across meteorologically sophisticated maps and imageries competing for space with quaintly primitive thunderbolts, smiling suns, and raindrops. Even here, however, the hyperreality of 3-D weather animations is superseding all other modes of accounting for the weather (Monmonier 1999, 181–89).

The Art and Science of Mastering the Weather

Notwithstanding its pointillist appearance and the illusion of perfectly aligned weather measurements that it perpetuates, the NWS forecast is a true product of collage—more, and different, than the sum of its parts. Despite their best efforts, weather forecasters are not able to master the indeterminacy of the atmosphere by the laboratory method alone, as Latour advocates. Instead, in order to precisely "make their predictions indisputable, to render the passage through their weather stations obligatory for everyone who wants to know the weather" (Latour 1987, 181–82), they have developed an omnivorous appetite for information and improvised for themselves a liminal epistemic space between the laboratory and the field, between the high-tech environment of the forecast office and the weather in the wild.

We have followed them in their adaptive collection and use of information as they try, suspended between place and a collaged space, to ground their pronouncements in an always elusive reality. NWS forecasters' information bricolage pursues weather observations of varying degrees of accuracy and sophistication in the name of a higher truth—ground truth. As such, it draws attention to the systematic and the messy, the standardized and the improvisational, the nomological and the resourceful, the deliberate and the tacit that makes up forecasting life on the operations decks of the NWS. On the one hand, it represents an organizational effort to project an exaggerated image of scientific rigor. On the other, it celebrates the art and science of mastering uncertainty. The end result is a hybrid, crafted out of a series of weather registers that not only include different and differing weather representations but also weather in the wild as well. There is nature in the mix that, although already domesticated into "nature," has nonetheless had a unique and overpowering influence on the overall meteorological composition. And while it never remains undomesticated for long, it is still relentlessly, undeniably there: past the door, on the roof, through the window, even inside the operations deck, forcing forecasters to adjust the blinds or the thermostat.

Effectively, forecasters' quest for the provisional coherence of the total observation speaks to the skillful creativity and discretionary judgment that go into producing the weather forecast. It foregrounds forecasters as artisans and weather forecast offices as craft guilds where one must first apprentice in order to acquire the necessary trade secrets and techniques to one day master atmospheric indeterminacy. It suggests that a masterful weather forecast is not reducible to computation and that forecasters must perforce move beyond mechanistic and prescriptive modes of seeing and interpreting the weather to fashion a compelling rendering of its future. As far as NWS forecasters are concerned, the brute force of a machine will always be stumped by the exquisite complexity of the atmosphere. A "feel" for the weather is mandatory if one hopes to prevail over meteorological uncertainty. Consequently, it would be a strategic mistake to allow oneself to become lulled by the infectious sense of control over nature permeating the NWS operations deck. To stay in touch with one's craft, one must have simultaneous access to a center of meteorological calculation and to "the real thing" outside. It is this direct sensory access to the ground truth of the field that makes the idea of weather forecasting as computer-aided human artistry palatable and marketable to the NWS administration.

That the best that they can hope to achieve is a collage—a provisional representation of reality, full of tensions and incongruities, that endeavors to rise above the sum of its parts—forecasters do not see as a sign of vulnerability or failure but as a challenge, a testament to their determination to improve on model-generated forecasts. Part of an organizational culture that actively promotes disciplined improvisation as an adaptive response to the constant state of ambiguity and uncertainty governing the decision-making task, NWS forecasters are primed to pursue complication and dissonance while all the while deeply engaged in pattern recognition and gestalt-building. In this they are hardly alone. The cobbling together of heterogeneous information as the art of collage readily brings to mind decision-making practices in a host of other settings characterized by a subjunctive mood and controlled chaos—from complex organizations to technoscientific collaboration to disaster mitigation. Drawing attention to the role of screenwork and digital compositing as the contemporary medium for gelling together into coherence resourcefully mobilized but otherwise disparate information cues, the heuristic of collage elaborates an increasingly prominent model of decision making in the digital age.

FOUR

Managing Risk: The Trials and Tribulations of Hazardous Weather Forecasting

Forecasting the Neborough weather is not for the faint of heart. The weather here showcases the best and the worst of all four seasons, and it is renowned for its fickle and dramatic personality. As one television meteorologist told me once, "Neborough weather is extremely complicated. People who come to forecast for Neborough come here to stay. Otherwise it's not worth it." Despite the range of Neborough's meteorological landscape, however, it is winter weather that dominates, stretching from as early as mid-October to as late as mid-April. This is the time of year when warm and moist low-pressure systems in the North Atlantic collide with arctic high-pressure systems from Canada to bring about the so-called Nor'easter storms the region is so famous for. When aligned just right, a Nor'easter will produce heavy amounts of rain and snow, hurricane-force winds, and high surf with severe coastal flooding and erosion. Not surprisingly, therefore, it is winter weather that commands the hazardous weather forecasting reality of the Neborough office and captures the imagination of its forecasters. To be sure, in the summer, the slightest indication that a hurricane may spin toward the Neborough coast has forecasters ruminating for hours over the latest hurricane model runs faxed over by the Tropical Prediction Center. In the summer, the thrill at the sight of a rotating, tornadic supercell on the radar and the tension over whether it is going to touch ground or not are so palpable on the operations deck you could cut them with a knife. Still, tornadoes, let alone hurricanes, are so-called low-frequency–high-impact weather events, especially for the Neborough region. Notwithstanding its potential for death and destruction and hence enormous meteorological excitement, the summer season is therefore generally considered a slow time at the Neborough office. It is the winter months—when the office, along with the atmosphere, is operating in "DJF (December/January/February) mode"—that forecasters can rely on to be their busiest, most stimulating time of the year. It is thus during the winter months at Neborough that one can best capture hazardous weather forecasting in action.

Hazardous weather forecasting encapsulates the essence of NWS forecasting, both in terms of its culture and its practice. It epitomizes the primary directive of the NWS to protect life and property and therefore carries with it additional organizational performance pressures. And it brings out the best in forecasters precisely because of the exquisitely complex challenges it presents. All the more so since it is exclusively the NWS that can legitimately issue weather warnings for the United States. Indeed, even if not always meteorologically challenging, hazardous weather still always poses a significant *forecasting* challenge thanks to its extraordinary public impact potential. As such, hazardous weather forecasting lends itself particularly well to a closer look at how the NWS and its forecasters negotiate a good forecast. It is here that one truly sees the processual, unfolding character of meteorological decision making come to life. And it is also here that it becomes clear that, while decision making in action is always prospective, risk management in action is often retrospective, resting on post hoc attributions to a single decision—or a single set of decisions—perceived to most credibly account for the given outcome. Risk management in action, then, is just as much a matter of making the right decisions as it is a matter of managing the fallout from what turn out to be wrong ones.

Cloudy with a Chance of Hazardous Weather

The word *weather* has several meanings in the world of meteorology. There is the generic meaning of the term, encompassing the state of the whole atmosphere. There is the colloquial meaning, comprising all surface weather phenomena and tantamount to what meteorologists refer to as sensible weather. Finally, there is the properly meteorological meaning of the term, standing for nondry weather, as in the phrase "There is no weather today."[1] In this last sense, the weather is now truly materialized, its tangible presence commanding the greatest attention because of the great risk to life and property it presents.

Similarly, the term *severe weather* has several meanings. In daily parlance, it is used interchangeably with *hazardous weather*, the assumption being that to be dangerous, weather must also be intense, or severe. Technically, however, the term *severe weather* is exclusively reserved for hail, localized—that is, "convective"—winds, or tornadoes. The NWS will thus issue a Severe Weather Statement only regarding an impending tornado or severe thunderstorm, the latter defined as severe if it produces "hail one inch in diameter (U.S. quarter-size) or larger, convective winds of 50 kts (58 mph) or greater and/or tornadoes" (NWS 2010). Yet this calculus of severe weather by no means exhausts what officially counts as a weather hazard. "Hazardous weather" also includes winter weather, and it includes nonprecipitation weather, such as dense fog or frost. Indeed, this formal distinction between severe and hazardous weather is but one example of the uneasy acknowledgment by the NWS of the social foundations of meteorological risk.

In sharp contrast to what is cast as universal consensus over the criteria for a hazardous thunderstorm—hence the insouciant collapsing of the terms *hazardous* and *severe* in the case of thunderstorms—the NWS criteria for hazardous winter weather are geopolitically specific and vary considerably. This variation, in fact, one encounters not only across regions and forecast offices but within the jurisdiction of a single office as well. The Neborough office, for example, will typically issue a Winter Weather Advisory if there is expectation of at least three inches of snow and/or sleet in the next twelve hours. Except, that is, for the northernmost state in its area of forecasting responsibility. There, the minimum threshold has been set to four inches, to account for the fact that this state climatologically experiences higher amounts of wintery precipitation and has therefore built a higher tolerance—both culturally and infrastructurally—for what constitutes "a significant inconvenience" or "a threat to life and property," the general NWS criteria for issuing an advisory and a warning, respectively.

To be sure, there are as many sensitivities to snow as there are people. The NWS solution to define snow hazards according to geopolitical boundaries is thus as reasonable as it is arbitrary. But the same argument holds true for severe weather as well, including thunderstorms. Because the NWS has resolved to define only some thunderstorms as severe, yet all thunderstorms initiate lightning, lightning is absurdly not among the criteria of hazardous weather, a pet peeve of many a forecaster. In the same vein, the one-inch hail threshold in the severe thunderstorm definition is based on the minimum hail size needed to cause "significant damage" to common roofing and siding materials (Marshall et al. 2002; NWS 2009c). Try telling a farmer, however, that a half inch of hail is not severe or significantly damaging to her crops. While forecasters may poke fun at a report of "severe lightning" by a volunteer weather observer or get impatient with a severe thunderstorm report that falls below the warning criteria, the persistence, quantity, and earnestness of such reports serve to underscore the inescapable outlandishness and artificiality of any absolute demarcation between hazardous and nonhazardous weather, be it a snowstorm or a thunderstorm.

That weather constitutes the socially marked subset of atmospheric phenomena has been a basic undercurrent of this book. But this chapter, on the trials and tribulations of hazardous weather forecasting, brings the matter to the fore. If it is the case that atmospheric dynamics are articulated as the weather only to the extent that they are considered socially salient, then the demarcation of hazardous from nonhazardous weather best highlights the social parameters of meteorological decision making. This demarcation, and the political and epistemic asymmetries in meteorological attention it fosters (cf. Brekhus 1998), becomes further institutionalized in the ways the NWS evaluates the accuracy and timeliness of its predictions.

Of Accuracy and Verification

It is fair to say that, in the eyes of the NWS, the accuracy of its forecasts represents first and foremost a performance marker, an opportunity to establish measurable and objective outcome targets as is expected of a government agency. As such, the claim to forecast accuracy inevitably calls for the administration of rigorous assessment protocols; it demands easily digestible numbers that can be analyzed and transformed into hard facts about the world. The snow accumulation measured at Metrocity International Airport thus counts as the snow accumulation for Metrocity. And a thunderstorm counts as severe as long as there is one report of a single one-inch hailstone. The search after this positivist chimera—reflective of the prevalent scientific preoccupation with accountability and parsimony, especially as it regards government spending—seeks to distill NWS forecasting practice into a performance score, effectively making the pursuit of accuracy synonymous with the pursuit of verification.

CHAPTER FOUR

NWS forecasters will readily admit that verification is a game, the pertinent question being not whether but to what extent to play along.[2] They are always mindful of the official verification criteria, and they are eager to meet NWS performance goals, taking extra care when forecasting for the known verification sites, as we saw in chapter 2, or being less proactive about issuing severe weather warnings for remote locations for which it is going to be next to impossible to obtain verifying reports. Still, first and foremost, verification is an *organizational* game. In contrast to the negative consequences that can befall the agency or an office should they not meet their stated verification targets, a forecaster's verification score does not seem to have an obvious bearing on his individual performance evaluation because, as a matter of NWS policy, it is not considered an appropriate criterion for assessing personal forecasting skill.[3] The logic and rules of the game, formulated at the level of the NWS administration, are consequently translated in complicated and conflicting ways at the level of forecasting practice.

After all, NWS forecasters are ultimately concerned with how to improve the accuracy of their forecast, not how to improve the accuracy of forecast verification. Whereas the NWS administration is endeavoring to make verification tantamount to accuracy, their preoccupation with accuracy largely begins where the organizational pursuit of verification ends. Characteristically, none of the forecasters at Neborough consult with any regularity the rather elaborate NWS archives of computer-generated verification scores. As far as they are concerned, the NWS verification scheme yields a skewed and misleading reconstruction of ground truth because it is point based, essentially confining itself to measures of temperature and precipitation at the handful of NWS automated weather station sites in the Neborough region. It may provide valuable feedback regarding these two weather elements at these fourteen locations,[4] but it has nothing to say about the rest of the forecast, which claims accuracy at a 2.5-square-kilometer resolution. When Neborough forecasters do consult these verification statistics, it is exclusively to gauge trends in their skill score, not to assess the accuracy of their forecast.

Quarterly office verification scores consistently meeting or exceeding NWS performance targets are ceremoniously displayed on the walls of Neborough's operations deck. Yet, whenever I asked Neborough forecasters to elaborate on their definition of a good weather forecast, they would make their point by precisely drawing a contrast with so-called objective verification. Echoing Fine's (2007, 183–84) informants at the Chicago office, they offered keen insight into their struggle to reconcile the pursuit of verification with the pursuit of accuracy:

It's hard because you got guidance to beat. I mean, you have to have healthy respect for the verification. But then you also have to keep in mind the service you are providing, too. Like today is a good example. The guidance yesterday was sixty or seventy percent chance of rain. Okay, so, whatever, I went with "likely." So, you know, numberswise, that appears to be a good forecast because we got the rain, we measured. But was it accurate? In reality, how much rain did we get? We got maybe an hour's worth of rain. Most of the day was dry, and when it rained it was just like nuisance type rain. So "showers likely" sounds like kind of a washout, you know. So you have to be careful: How much of the day did it actually rain? Did it rain for five minutes and you measured? That could be deceiving. You have to look beyond the numbers, the numbers don't tell the whole thing is what I'm basically getting at.

Forecasters resist equating a verified forecast with an accurate forecast. A "healthy respect" for the numbers is necessary but, ultimately, it is service that counts. The invocation of public service—the attribute par excellence of a good NWS forecast—keeps verification and accuracy juxtaposed, at an awkward distance from each other: verification concerns dictate that one play "the numbers game"; accuracy concerns dictate that one strive to look beyond the numbers to what "actually" transpired.

The threat of hazardous weather largely calibrates out the tension between verification and accuracy. At the administrative level, the sociopolitical freight attached to a missed hazardous weather warning militates against a strict adherence to objective verification measures and in favor of a more adaptive approach that also incorporates human reports, not just observations from automated weather stations. The threat to public safety forces the NWS to acknowledge, and attend to, the local variability of weather and the vagaries of its measurement.[5] Hazardous weather reports are compiled and quality checked on the ground by forecast offices, which are also tasked with tabulating and reporting back quarterly and seasonal verification tallies to the regional headquarters. To be sure, despite recent efforts to standardize local verification procedures, hazardous weather verification remains controversial because of its potential for high levels of subjectivity and instances of conflict of interest (e.g., U.S. Department of Commerce, Office of Inspector General 1998, 12ff.). But the weight of the situation gives administrators license to shift the focus to tangible forecasting "hits" instead of forecasting rigor.[6]

For their part, forecasters feel compelled by the weight of the situation to closely adhere to the performance measurement protocols set by the administration. At stake now is protecting life and property, not beating the models. Neborough forecasters thus keep meticulous records on every hazardous weather event, warned or missed, filing together in pa-

CHAPTER FOUR

per folders all forecast products issued, all pertinent weather and damage logs and reports, along with the derived verification statistics. Even after a long shift of trying to keep on top of the weather, they make sure to properly archive all relevant information. And it is a testament that they are prompted to this type of painstaking bookkeeping not simply because they must follow protocol that verification folders remain on the operations deck several weeks after for easy reference and study. Under the threat of hazardous weather, the demands of verification and accuracy converge and NWS administrators and forecasters find common ground. Exceptionally, forecasting skill is officially measured against weather reports, not model performance. Exceptionally, forecasters may speak interchangeably of ground truth and verification data.

Indeed, in practice, hazardous weather verification is seamlessly integrated into the forecasting task. During a forecasting episode certainly, it becomes unrecognizable as such, organically blending into the standard improvisational process of maintaining situational awareness (cf. Murphy and Winkler 1992). But even afterwards, when forecasters are on the hunt after reports for any unverified warnings and the temptation to indulge in the verification game runs high, the motivation to understand and learn from what just transpired can be just as strong, albeit biased toward unusual meteorological scenarios. Thus, when the Neborough office acquired the technology to run simulations of locally archived weather data in October 2003, a "Post-Storm Analysis Team" was immediately formed to study "significant or unique weather events" so as to "help address the decision making that occurred for better or for worse."

Getting the Word Out

In organizational terms, the essence of a good NWS forecast is currently distilled into two metrics: accuracy and timeliness (NWS 2011b). Extraordinary times beget extraordinary measures, however. If the threat to public safety compels the NWS to asymmetrically attend to the accuracy of its predictions during a hazardous weather episode, this is all the more the case when it comes to the timeliness of its hazardous weather predictions. Under the threat of public safety, in fact, getting the word out usually trumps most any accuracy qualms:

Intent on figuring out "why I lost all my snow," Cheryl came in two hours early today to spend time with the archived data of yesterday's storm. Soon, Simon and Wayne joined the "post mortem" in progress. . . . Wayne gets pretty excited. Keeps coming back to the screens while meaning to leave, has Cheryl bring up more radar data, pull up infrared satellite imagery, to try and follow the sleet line. The meso-NAM low level "masks" what happened on the atmosphere higher up, he says frustrated, "it's deceptive." "It was all there in the upper air sounding, but with everything else going on we didn't have time to look at it!" . . . From a verification standpoint, the office did an excellent job because they reached the warning criteria of seven inches. But verification scores, says Cheryl, "are not very nuanced"—"we busted the forecast by five inches." . . . Simon, however, is more sanguine about what transpired: "This is snowband behavior. Of course we're not going to nail it, but overall we did a pretty good job with the timing of the snow yesterday. . . . Seven inches, twelve inches, it doesn't really matter. They were waiting for it, that's the important thing." "The nicest feedback I ever got," he says, came from the doctor two doors down where he took his daughter who had bronchitis yesterday: "You guys got it right both times. More than one foot last time. You said this time the snow was going to start at 2:00 p.m., and at 2:30 p.m. my wife called to say my son's game had ended because of the snow. And you said it would stop at 6:00, and it turned to sleet at 6:00. I looked at my watch." . . . "We got the word out," Simon kept saying wisely throughout the shift, "that's what matters."

In order "to increase public awareness and promote a proper response" (NWS 2013b), the NWS has instituted a staggered warning communication system, which begins with a Hazardous Weather Outlook, escalates to a Watch, and, depending on the severity of the situation, can result in an Advisory or a Warning.[7] The Outlook is issued up to seven days ahead of time to highlight a potential weather threat if there is at least a 30 percent probability of occurrence. Up to forty-eight hours ahead of time, the Outlook is upgraded to a Watch if the probability of occurrence has increased to at least 50 percent. And up to thirty-six hours ahead of time, the Watch is upgraded to an Advisory or a Warning if the probability of occurrence has increased to at least 80 percent. Meanwhile, so-called "nowcasts" and "special weather statements" are to be issued in frequent intervals during, or in anticipation of, a given hazardous weather event to reinforce and amplify the NWS message, provide updated information about existing and impending conditions, and outline appropriate response actions.

Yet, while this staggered warning system has ostensibly been implemented by the NWS for the purposes of disseminating to the nation critical weather information in a timely manner, it also doubles as a safeguard against the excesses that hazardous weather can inspire in its forecasters. Reigning in the conflicting pull between serving the science and serving the public presents a formidable challenge under routine circumstances. With public safety at stake, forecasters' delicate balancing act acquires

CHAPTER FOUR

a new moral dimension, a new urgency, as it becomes compressed into the question, to warn or not to warn? When transformed into a matter of overwarning or underwarning, accuracy concerns may be silenced in the name of public service. Just as likely, the eagerness to "nail the storm" may cloud judgment as to how and when to get the message out. By requiring a 30, 50, or 80 percent confidence level before a forecaster can issue a hazardous weather outlook, watch, or warning, respectively, the NWS aims to curtail decisional flip-flopping and encourage consistency and continuity across forecast shifts and offices.

And there is an ever-increasing number of further checks to ensure that the NWS is speaking in one voice. A recent NWS policy, for example, mandates that a winter weather watch can only be issued if at least one adjacent office agrees to do the same. In the summer, when performance pressures on the NWS reach a climax given the devastating potential of a hurricane or a tornado, the administrative impetus to regulate forecasting practice intensifies further, exacting consistency and timeliness at the national, not simply the regional, level. Offices are expected to initially abide by the severe weather watches issued by the Storm Prediction Center, while, during the hurricane season, they are constrained by the tropical storm tracks charted by the National Hurricane Center and, in fact, are expressly forbidden from publicly divulging any information before its official release by the Hurricane Center in order to guarantee "the issuance of information to all users at the same time on an equal basis" (NWS 2013a, 3).

In sum, the institutionalization of the NWS "early warning" weather system presupposes the institutionalization of multiple other operational policies across distinct organizational fronts. More than streamlining the warning dissemination process or the coordination with emergency management and state officials, getting the message out in a timely manner hinges on the production of a hazardous weather forecast that, first and foremost, is reliably consistent. Hence the redoubled efforts by the NWS administration to articulate and enforce explicit forecast collaboration policies during hazardous weather operations. Even as it justifies such extraordinary centripetal pressures toward a unified decision-making approach, however, the threat to public safety excites equally strong centrifugal tendencies toward a locally sensitive response to the potential for hazardous weather. Fine (2007, 146–52), who spent two weeks observing operations at the Severe Prediction Center (SPC), nicely captures the obduracy of this tension in the negotiation tactics SPC has come to develop in order to mold the risk assessments of individual field offices into a single coherent severe weather watch. And I had occasion to witness a number of less than amicable conference calls between Neborough forecasters and their counterparts at SPC or the Hurricane Center over the track and intensity of a particular storm system. At times, such altercations may lead to productive, collectively derived outcomes. In 2005, for example, a frequent point of contention between SPC and weather forecast offices, the infamous severe weather "watch box," was happily resolved in favor of designating potential watch areas by a list of counties instead of an unwieldy parallelogram. Yet this center versus field tension is unsolvable at its core, itself an instance of the broader dialectics between reductionism and emergence that keep fractally regenerating as one moves further inward—from NWS headquarters to NWS offices to NWS forecasters. Extraordinary times only make the challenge to reach a decision that steers clear of either extreme that much more difficult.

Busting the Forecast

In an effort to throw meteorological decision making into the highest relief, the balance of this chapter is structured around two missed weather events. The first incident illustrates the fragile dance between accuracy and consistency and the serious repercussions that may arise from a missed hazardous weather forecast in the absence of any hazardous weather whatsoever. The second incident details the aftermath of a forecast that missed the first snow of the season. Taking place under what are considered to be "typical" and "exceptional" hazardous weather forecasting conditions, respectively, these episodes progressively excavate the coconstitution of the social and the atmospheric in weather forecasting operations.

Where is the Snow?!

By all accounts, the 2003/4 Neborough winter proved rather uneventful weatherwise, especially when compared with the winters immediately preceding and following it. This is not to say that things did not get off to a promising start: the first major snow event of the season was a major snow event indeed, generating up to thirty-five inches of snow around the Neborough region and eighteen inches in Metrocity alone.[8] Alas, arriving during the first week of December, atypically early for a storm of such magnitude and ferocity, it set the stage for a winter that never really came. None of the big snowstorms slated for the following weeks materialized. Contrary to the hopes and expectations of the Neborough of-

CHAPTER FOUR

fice, December 2003 was not meant to be the fifth snowiest December in Neborough climatological history, after all. By the third week of January, with the weather and the models continuing to tease them, forecasters' nerves were visibly frayed. The models were hinting once again that "it's going to get very interesting" in a week's time. With his usual flair, Wayne wrote in the Area Forecast Discussion,

Ready-set-go on moderate or greater winter events next two weeks. Time has arrived for a reality check on the expectation of above normal wintry precipitation. Suggest to be prepared for large amounts of snow and/or ice . . . Not all models favor big next two weeks, but GFS ensembles itself continues strong and believe the pattern favors higher than climatological risk for above normal wintry weather.

On January 25, the Neborough office, in collaboration with surrounding offices, issued a strongly worded winter weather outlook, alerting to the potential of a foot or more of snow for the entire region. Despite some timing differences among the models, the heaviest snow was slated to occur between midnight and 7:00 a.m. on Wednesday, January 28. By early morning Tuesday, with models continuing to converge on an overnight burst of heavy snowfall, the winter weather watches had been upgraded to winter storm warnings, which alerted to the potential for eight to as much as sixteen inches of snow for parts of the Neborough region. The 12Z (7:00 a.m. Eastern Standard Time)[9] model runs later that day, however, were all singing a different tune, keeping the shallow coastal low-pressure storm system separate from the upper low and thus advising against a widespread heavy snowstorm. Neborough's reaction to the new model trend was calm and diplomatic as reflected in the afternoon Area Forecast Discussion:

12Z GFS and short range ensembles support our earlier suspicions that the snowfall forecast may be a tad high, and this falls in line with what we have seen on 12Z ETA, so we will back off a little on total accumulations. As a result we [will be] going with 3–6 inches in northern Neborough, 4–8 inches along the Interstate, and 6–10 inches in southern Neborough.

All warnings were kept in effect, albeit with decreased snow amounts, and the watch for northern Neborough was converted into an advisory.

At 6:30 p.m., all was quiet on the Neborough operations deck. The evening shift, consisting of Margaret and Arthur, had been augmented by Peter, primarily there to man the phones. The snow was not expected to arrive before 10:00 p.m., and the media calls—bound to resume just in time for the evening news—had let up for a while. This quiet, however, was pregnant with apprehension about the continuing downward trend of snow accumulation from the models, now buttressed by "reality" from the satellite imagery and radar scans from the neighboring NWS office. In the midst of furiously dissecting her screens, Margaret called out to Peter, also engrossed in his screens over at the hazardous weather desk:

Margaret: Hey Peter, you noticed how closely the [radar scan] matches the 12Z RUC and meso-ETA?
Peter: Yes, and they don't show any snow north of the Interstate.
Margaret: We may need to back it down.
Peter: Yeah. Take a look and let me know what you think.
Margaret, *a minute later*: The RUC is coming drier than before.
Peter: Yup.
Margaret: It's scary. [Northern Neborough] is getting nary a flurry.

Satisfied with her assessment of the latest data, Margaret takes out a fresh copy of the Neborough county map and draws in her predicted snow amounts, referring to the map of the previous shift as she does so. Map in hand, she wheels herself over to Peter, who has his own map ready. They seem genuinely dumbfounded by how well their advisory/warning cutoff line matches up. Their snow amounts differ slightly, of course, and Margaret included Metrocity in the higher snow amount zone, "just to be on the safe side"—but they regard these differences as minor, "splitting hairs." Together, they come up with a new warning map to serve as the basis for all subsequent forecast updates during the shift. Essentially, they decide to take two inches off the maximum snow accumulation for all the warning zones north of the interstate. This in turn means that they will have to cancel the advisory in northern Neborough and downgrade all warnings north of the interstate, including the warning for Metrocity, to advisories for snowfall amounts in the three to six inches range.

Despite the difficult position they found themselves in, Margaret and Peter were careful to maintain consistency with earlier forecasts, both by consulting the previous warning map and by gradually cutting down snowfall amounts and downgrading warnings. Indeed, the remarkable similarity between their ostensibly independently derived warning maps is not that surprising, after all. By keeping in step with prescribed protocols, they were bound to lock into the one, organizationally sanctioned, forecast solution. When the institutional and procedural logic of decision making privileges consistency over accuracy, as is especially the case during hazardous weather forecasting, then the right answer presents

CHAPTER FOUR

itself as the only answer. Not always, as we shall see in a moment, but certainly more often than not. What in reality is a messy, open-ended, dynamically evolving *process* of decision making—by multiple forecasters, in the course of a single shift or many shifts, according to the latest weather reports, in consultation with the latest meteorological guidance, as a result of various deliberations—may thus be neatly collapsed into a single decision, or a set of decisions, after the fact. What in reality is a verb, forecasting, thus becomes a definite noun phrase: *the* forecast. To be sure, process and structure recursively inform each other. Both verb-based and noun-based understandings of decision making are therefore necessary for organizing NWS forecasting (cf. Weick 1995; Bakken and Hernes 2006). Yet, even under the most routine fair-weather conditions, when forecasters may be inclined to see the decision-making process as obvious and straightforward, a decision already made, they still cannot but remain attuned to the evolving empirical situation at hand. Routine forecasting is no more passive and programmed than nonroutine forecasting is creative and nonprogrammed.

By 10:30 p.m., with the midnight shift in already, everyone on deck was anxiously studying the radar, willing the spotter phone lines to ring. In the absence of any actual reports from the ground, the radar was treated as only an indicator of snowfall, and a deceptive one at that.

Margaret: Now, according to the [radar], it's supposed to be snowing all the way up to X-town.
Dick: Well, yesterday, according to the radar, it was supposed to be snowing whole chunks over Y-town, but was it?
Margaret: That's because it was too dry . . . too dry . . .
Dick: Arnold, do you have a timing on this thing? We should issue a nowcast.
Arnold: I don't know. That's what I'm trying to do, but the radar is blowing me off . . .
Margaret: I have eliminated some of the background noise, and I think I have an approximate time. I'm picking between 1:00 and 2:00 a.m. in B-town and in C-town. And I think it will reach Metrocity around 4:00. Maybe a little earlier, maybe a little later . . .

At 12:46 a.m., the first spotter call came in from A-town, reporting three inches in one hour. A couple more calls followed, but otherwise a deafening silence reigned over the operations deck. As the hours ticked away, the midnight shift became increasingly nervous and tense, obsessively checking model data against the ground truth information available. As hard as it was for them to recover from the fact that "what I was looking at last night and what I'm looking at today are completely different," all hope was not lost that Metrocity, at least, would reach advisory criteria, as forecast. The nowcast issued at 3:09 a.m. projected a deceptive air of confidence: "A band of snow will . . . reach Metrocity between 4 and 5 am. Once the snow begins it will quickly become heavy with snowfall rates of 1 to 2 inches an hour for about 2 hours. Driving conditions will deteriorate quickly once the snow starts making for a very slippery morning commute."

Around 3:30 a.m., my exhaustion amplified by the palpable dejection on the operations deck, I decided to briefly retire to the storeroom, which also served as makeshift sleeping quarters for the staff. I woke up at 7:00 a.m. to find the midnight shift gone and Ray in a foul mood at the short-term desk, busy "cleaning up the mess of the previous forecast." Apparently, the midnight shift had decided to cancel *all* the winter weather warnings and advisories out at 4:20 a.m. because they did not receive any snowfall reports for Metrocity. Once the phone calls started coming in around 5:00 a.m. from other parts of the region, they issued a nowcast hoping this would be enough to control the damage. Alas, as the snow and the spotter reports persisted, Ray was forced to issue a new advisory for south Neborough instead of just downgrading to one, as he would have liked, and then expand the advisory to additional counties two hours later. Meanwhile, the day crew slowly trickled in, good-naturedly commenting on the situation: "Hey Ray! Missed us, man! I woke up this morning, looked outside and said, 'What?!' . . . I heard X-town got hit, though." To me: "Ah, so you were here when they started busting: 'Oh, no! Where is the snow?!'"

Ray was both enraged and mystified about "what possessed them to pull the plug": "It makes us look stupid. We looked stupid before for having predicted more snow than we got, but we look more stupid now for flip-flopping." Up until the evening shift, he continued, the office was backing down the snow amounts in a smooth fashion. But the midnight shift lost faith too quickly. Because the snow did not come by 4:00 a.m. in Metrocity, they thought it was not coming at all. "If we have a warning out," Tom lectured helpfully over at the long-term desk, "we have to be really sure before taking it down, and usually you want to downgrade. One must gracefully step back; or else keep the warning, no more harm done." It is true that the model runs looked very different from 12Z to 00Z, he conceded, "but there are two schools of thought about models: one takes them literally and the other takes them as guidance. The mid[night] shift should have known better."

To clarify: the decision to cancel the winter weather advisory for Metrocity was, by all accounts, a sound one. As it turned out, Metrocity only

CHAPTER FOUR

received one-and-a-half inches of snow. In a frantic attempt to save face, however, forecasters dispensed with forecasting etiquette and canceled all other warnings and advisories still out for the Neborough region. Given that the models indicated the atmosphere was too dry for further snow development, the absence of snowfall reports from Metrocity signaled the death knell for the entire forecast. The drama on the operations deck is captured in the Area Forecast Discussion that followed the cancellation:

Little hope for any additional moistening of the air mass. Model soundings show that best upward vertical velocity north of the interstate is placed above best snow crystal growth. (Interestingly enough, this was depicted and noted last night but with other dynamics below this, focused on that.) But never mind models at this point, have to go with ground truth information. And with that, will take down warnings.

Although the decision of the midnight shift to cancel the other warnings proved to be wrong in hindsight, maintaining consistency and "stepping back gracefully" can have just as deleterious a public effect, however. The tendency, among decision-making practitioners and scholars alike, to ex post facto chart a single linear causal path from decision making to its outcome obscures the fact that in reality a given good/bad outcome often emerges out of multiple and conflicting right/wrong decisions. What is signaled as *the* fateful decision after the fact cannot but be based on incomplete, if not downright arbitrary, considerations. To paraphrase Weick (1995, 184–85), it is an act of interpretation: it "consists of locating, articulating, and ratifying [some] earlier choice, bringing it forward to the present, and claiming it as the decision that has just been made." In this case, did Neborough forecasters overforecast or underforecast the storm? And which one should be considered the critically wrong decision? There is a strong argument to be made that overwarning is the lesser of two evils. But one need not look further than this very episode to appreciate the serious repercussions that may arise from a missed hazardous forecast even in the absence of any hazardous weather whatsoever.

Per Metrocity policy, whenever the NWS predicts four or more inches of snow, the city is under a snow emergency, and residents have four hours to remove their vehicles from the streets. Recall that up until the eve of the storm, Tuesday the 27th, the Neborough office had a warning out for Metrocity calling for four to eight inches of snow—down from the six to twelve inches advertised in the morning's watch. At 7:00 p.m. that evening, the newly appointed mayor of a Metrocity town declared a snow emergency, and police cars roamed the streets until 11:00 p.m., announcing with megaphones that designated routes had to be cleared of parked cars, or such cars would be ticketed and towed. Around 7:50 p.m., over at the Neborough office, Margaret and Peter downgraded the warning to an advisory, intent on maintaining consistency, and called for three to six inches of snow. Meanwhile, even as they were dutifully broadcasting the NWS advisory, meteorologists at several television stations were predicting one to three inches of snow for Metrocity instead. In the following hours, with nary a flake in sight, three thousand cars in Metrocity were ticketed and another two hundred towed before the snow emergency was lifted at 4:20 a.m., right after the Neborough office canceled the winter weather advisory. The next morning, Metrocity residents woke up to one and a half inches of snow, per the official measurements at Metrocity International Airport. When many of them found—or did not find, as the case may be—the parking violation tickets, a small riot ensued. According to the Associated Press, which picked up the story, the mayor had no plans of forgiving the $180,000 of parking and towing fees. "We are at the whim of Mother Nature here," he said. "We are at the whim of projections and forecasts. Sometimes it may snow more, sometimes it may snow less, sometimes not at all. We're all in the same predicament as far as projections." He claimed that his decision to ticket and tow was made, in part, to avoid a repeat of what happened after the big early December storm, when poor enforcement of snow-emergency procedures—by the former mayor—resulted in clogged streets. "Public safety is the issue at the forefront. . . . We ask that residents work and cooperate with the city to help achieve our public safety goals and get the streets clean and clear." One day later, however, with the story getting increased attention from media across the region, the mayor announced he was granting a "onetime amnesty offer" on the charges. Looking back, he admitted, "I think I have never prayed so much for snow in my life."[10]

Back at the Neborough office, forecasters were also praying for snow. With the advisories out, everyone on deck was now furiously looking for snowfall reports to verify the earlier warnings. The minimum snow amount required was an average of six inches per affected county. Running down the list of registered weather spotters, they located someone in a coastal county with no snow reports as of yet. Ray hung up the phone: "He said three inches." Peter: "Three inches? Liquid? That doesn't sound right, it should be more than that." Next up was the harbormaster, who reported six inches. "That's more like it! He's the harbormaster, he knows what he's doing." Upon which Ray suddenly remembered that the previous spotter had said it was very windy, and so the snow must have been drifting, which would explain the low report, he reasoned. He accordingly logged *two* reports of six inches of snow, thereby verifying the warning for that county.[11]

CHAPTER FOUR

Later that same day, I observed Ray compile the verification statistics for the just transpired storm. He first consulted observation data to determine when each Neborough county was bound to have reached advisory or warning criteria. With the help of the already quality-checked final snowfall reports, he then derived the average snowfall total for each county, recording the amount on a Neborough warning map and interpolating snowfall values for counties with missing reports. Finally, he calculated the required statistics. While all the advisories were issued after the fact and therefore counted as missed events, Ray judged that five out of the thirty-five warnings issued the previous day had verified before they were canceled by the midnight shift. According to Ray, and he had to justify his decision repeatedly in front of skeptical members of the staff, these counties had met the six-inch warning threshold before 4:20 a.m. based on the timing of the snow reports in some of these counties and the overall track of the storm. All in all, Ray deemed, "verificationwise, we didn't do too bad really, except for Metrocity, and that's the big media capital. We got hammered there."

Despite the forecasting debacle of that day, Neborough verification scores for the 2003/4 winter season managed to meet and in fact slightly exceed the verification target goals set by Eastern Region Headquarters. But that should not be very surprising. Like all NWS forecasters, Neborough forecasters are quite savvy players of the verification game and of its ambivalence as both decision-making and impression-management tool. Truly, they know when to hold them, and they know when to fold them. And should they find themselves in a sticky forecasting situation, the NWS directives themselves provide detailed case-by-case examples of how best to proceed so as to preserve "customer confidence." That is why the actions of the midnight shift appeared so mystifying and inconceivable to the rest of the staff in the light of day. Time and time again, the discussion on deck kept returning to what went wrong. As is ordinarily the case for extraordinary weather events, staff with no immediate forecasting duties clustered around the screens of the Warning Event Simulator scrutinizing the model and observation data of the past seventy-two hours, already archived by Peter for in-depth analysis later. "Look at 21 dBZ [decibels of radar reflectivity] not making it to the ground. Wow!" "Snowed like hell at two thousand feet, but the temps were warm enough that the snow was evaporating away before it reached the ground." "Well, consider looking at that on the radar. Why not drop the warning?!"

In the next few days, the office's Post-Storm Analysis Team wrote up a four-page report of the event, subtitled "the Over-Forecast of the January 28th Storm." The paper and the relevant model and observation data sets were uploaded to the NWS Case Study Library and made available to other offices for training purposes. Importantly, while the analysis otherwise includes a detailed discussion of decisions made, there is no mention of the decisions of the midnight shift. When queried, the authors insisted their case study was about the overforecasting of the storm; the flip-flopping of the midnight shift was "not meteorologically relevant" and therefore not appropriate training material. Yet this stance must be at least partly attributed to their well-cultivated disposition to "save face" for the office and the NWS at large. Overforecasting can be made palatable and turned into a learning opportunity by evenly distributing the blame across a broad network of actors, including the atmosphere, computer models, and observation instruments (cf. Fine 2007, 44). Not so with flip-flopping. It constitutes too grave a breach of meteorological conduct, according to the logic of NWS forecasting, to be similarly diffused. By excluding it from the narrative, therefore, the Post-Storm Analysis Team preserved the integrity of the report. This was still a *wrong* forecast—a "storm bust," per the report's original subtitle—because it ended up overforecasting the winter storm. But it no longer exhibited traces of *bad* forecasting. In the report, which accounts for the movements of each shift up until the midnight shift, Neborough's decision making appears carefully calibrated and steady, gradually downgrading the potential effect of the storm in response to the "subtle but significant changes" of the 12Z and later the 18Z model runs, and "in concert with its neighboring offices." Absent the drastically divergent, although at the time no less sound, forecasting decision of the midnight shift, the apparent principled consistency of Neborough's decision making lends it legitimacy: an "overforecast" in hindsight to be sure, but an eminently sound one given what was being "suggested" by model guidance at the time.

This is not to say that the strategy of the Post-Storm Analysis Team to isolate the decisions that contributed to the overforecasting of the storm is without merit. On the contrary, it helped them to identify "the missed clue"—namely, that model guidance had mishandled the evolution of the jet stream three days ahead of the storm—and to locate the same model performance pattern in an overforecast storm two months earlier. Still, to the extent that NWS forecasters have come to genuinely believe they can legitimately bracket out the social implications of their decision making,[12] so poignantly evoked during moments of flip-flopping, this is clear evidence of an ingrained obliviousness to the coconstitutive role of the atmospheric and the social in weather forecasting operations. Such obliviousness might be unfathomable to the rest of us on the outside, but as I discuss in the next section, it represents a reasonable institutional re-

CHAPTER FOUR

sponse to the impossible charge of protecting life and property from the vagaries of the atmosphere.

The First Snow of the Season

As one would expect of a government bureaucracy, the NWS has taken great care to operationalize notions of hazardous weather into a specific, albeit complex, set of standardized meteorological criteria. Such operationalizations, of which I outlined only a few in the beginning of this chapter, draw attention to the social underpinnings of meteorological decision making, and they help trace the linkages spanning the material and the social dimensions of the atmosphere. Still, it is the exception that makes the rule, and it is the exception to the typically well-defined criteria for identifying hazardous weather potential that best showcases the social stakes involved. One encounters such an exception in the NWS Directives for issuing winter weather warnings. Forecast offices are advised that they should issue a warning when there is an 80 percent or greater chance of a hazardous winter weather event meeting local warning criteria, *or for high-impact events which do not meet local warning criteria*. The text continues,

> The following is an example of impact vs strict criteria: Winter storm is forecasted but accumulations will not meet traditional criteria. However, if it is early in the season or during a critical time of day such as rush hour when the impact will likely be high, then a Winter Storm Warning might be warranted. The forecaster has the discretion and should not be held back from issuing what best describes the impending winter hazard even if traditional criteria may not be met in the strictest sense. (NWS 2013b, 6)

One page earlier, under the subheading "Forecaster Judgment," the same NWS Directive already acknowledges that

> written instructions cannot address every operational situation. All [weather forecasting office] personnel exercise initiative and professional judgment to minimize risk to public safety and property, constraint of travel and commerce, and needs of users in situations not explicitly covered by written instructions. Protection of life and property takes precedence in these decision-making processes. As such, criteria for winter storm warnings are considered guidance only, not strict thresholds. Forecasters may issue warnings and advisories based upon lower criteria if the event in question poses a significant threat to life due to timing or other circumstances. For example, an advisory or warning may be appropriate for a minor snowfall event that takes place near rush hour, even if the amount may not meet strict criteria. (NWS 2013b, 5)

In the same vein, the related Eastern Region Supplement stresses that "if costly transportation impacts occur, especially from the first snowstorm of the season, this can be counted as an [hazardous winter weather] event" (NWS 2005b, 3).

It bears noting that the NWS Directives do not stipulate a similar exemption for severe or nonprecipitation weather warnings. Consistent with the prevailing institutional logic that sees the public safety hazard posed by flood or dense fog as universal and the public safety hazard posed by snow as locally specific, it is supposedly only winter weather that occasionally requires a relaxing of the rules in order to protect life and property. *Exceptionally* exceptional, then, the organizational plea to disregard formal procedure in the face of the deep-rooted impulse to rationalize and standardize NWS forecasting practice promises important insight into the fundamental concerns that drive meteorological decision making.

The forecasting episode that serves as an illustration is particularly instructive in this respect. It details the aftermath of a forecast that missed the first snow of the 2003/4 winter season, a forecast that would most likely have gone unnoticed only a few days later. Indeed, the episode in question occurred just as forecasters were bracing themselves for the big early December snowstorm that would end up dumping eighteen inches of snow in Metrocity. Under the headline "Real Winter Making a Run for Neborough" and mostly concerned with the upcoming Nor'easter, the Area Forecast Discussion of the previous day made a brief mention of a reinforcing cold front that "may produce a quick burst of snow showers" between 2:00 and 7:00 a.m. tomorrow morning. "A few locations of the higher terrain could see a quick coating to one half inch of snow. As the front moves through, could see a few locations approach borderline wind advisory criteria." Accordingly, the forecast advertised a 30 percent chance of snow showers for some counties, including the greater Metrocity area. That night, however, the midnight shift downgraded the forecast for Metrocity from a chance of snow *showers* to a chance of snow *flurries*. Note that snow flurries involve light snowfall that generally leads to "just a dusting," whereas snow showers involve snowfall of varying intensity that typically leads to some snow accumulation. Snow *squalls*, meanwhile, are intense snow showers with moderate to heavy snow accumulation.

On my way to the office from Metrocity around 5:30 a.m. that morning, I noticed that cars with plates from the northernmost state in the Neborough region were shedding snow from their roof. Little did I suspect what was yet to come.

CHAPTER FOUR

6:40 a.m. When I arrived, John (long-term desk, midnight shift) greeted me with the question, "Did you have snow in Metrocity?" I said no but told him about the cars. He knew. They already had reports of up to two inches there. What he was worried about, though, was that the snow showers might make it to Metrocity. Their forecast only had flurries for Metrocity.

. . .

7:00 a.m. Margaret (short-term desk, midnight shift) is briefing Wayne (short-term desk, day shift) about the situation: "We were fine until 6:00 a.m., and then it started." She said they had "covered the situation but not perfectly."

. . .

7:20 a.m. The big news is that there actually was snow in Metrocity! Metrocity International Airport reported not snow flurries, not snow showers, but snow *squalls* between 6:45 and 7:10 a.m.! I missed it driving here. Meanwhile, the forecast was still advertising a chance of snow flurries. . . . Phil (long-term desk, morning shift) to Wayne: "How are you doing, Wayne?" Wayne shakes his head. Phil, chuckling: "Not good, eh?" Wayne: "Already updating the forecast. Not the way I would have . . ."

. . .

8:00 a.m. Margaret is leaving and passed by Wayne to pick up her stuff. "I'm issuing a wind advisory," Wayne says without looking up at her, "Metrocity Airport reports 39 mph." Margaret frowns. She collaborated with surrounding offices, she repeats somewhat defensively, and, except for [one], they had all agreed the winds would be below criteria.

It was only later in the morning, however, as calls from the media started flooding the office, that the significance of what had just transpired was fully grasped. Striking Metrocity during the early morning commute and reducing visibility to as little as a quarter of a mile, the squalls left behind a thin coat of snow and black ice along with more than three hundred accidents statewide, many involving jackknifed trailers that blocked interstate exits. While it was the northwest part of the region that saw the worst of the storm, nevertheless, the commute ground to a complete halt for over five hours on nearly every route into and around Metrocity. Road crews, urgently sent to the scene to spread salt and sand, got stuck in traffic alongside everybody else. No casualties were reported for Metrocity, but there were two fatalities and several injuries due to traffic accidents region-wide.

As is to be expected, the "surprise squalls" made national headlines: "It was one little inch of snow. It may as well have been two feet." "The squalls lasted a few moments as they stormed across the state. The morning commute lasted into the afternoon." The news stories featured a quote from the spokesman for the Highway Transportation Department saying that his agency had been caught off guard because the weather forecast had been for "mere flurries." After which, there was predictably a line from a Neborough forecaster, usually Wayne, who was quoted saying, "It wasn't so much a surprise for us. We had snow showers in the forecast."

Wayne's statement is technically true in every respect: the snow squalls were not so much a surprise, and/because the office—specifically Wayne himself during the previous day shift—did have snow showers in the forecast. So what if the showers had ultimately been downgraded to flurries by the next shift? So what if the intensity of the wind that transformed the snow showers into squalls had been a surprise even to Wayne? The processual, open-ended character of meteorological decision making allowed him to select as *the* Neborough forecast the one that best let him rhetorically manage the fallout from the missed squalls, the one that best made it "accountable" (Garfinkel 1967, vii) to himself and to others. I was there to witness his phone conversation with the Associated Press reporter, and I witnessed other, downright hectoring, media calls that day inquiring of Neborough staff, "Is it *that* hard to forecast snow squalls?" No matter. While they may occasionally fumble under the pressure, NWS forecasters would never blink in front of a reporter. For one, as scientists in the public domain, they have become experts at impression management. Equally importantly, however, their efforts to save face in public typically find a sympathetic audience with beat reporters. Routine news work, after all, serves to legitimize the status quo by actively repairing and filtering out incongruous information (Fishman 1982). The requisite "remedial interchanges" (Goffman 1971, 108–18) between forecasters and reporters toward a mutually acceptable redefinition of the situation, then, are usually conducted swiftly and efficiently. It is rather the newbie on the beat who might pursue a less than cordial line of questioning, prompting the forecaster taking the call to adopt the exaggeratedly calm and deliberate tone one uses with a petulant child. Either way, NWS forecasters take on the role of *explainer* of—not apologizer for—what transpired.

And so, once again, the Neborough office managed to keep up appearances, allowing the press to do likewise. Most stories adopted a human interest angle, while the rest focused on the ongoing heated contract negotiations between snow removal companies and the state. In fact, several of them concluded with an NWS sound bite about the upcoming snowstorm: today's squalls "might be setting the stage for something this weekend. People might start hearing talk of a mid-Atlantic storm, but it's too early to tell how intense it will be at this point." The authority of the NWS, such as it was, had been reasserted.

Negative examples can afford great analytic insight into the hidden assumptions and processes underlying a phenomenon under study. That,

CHAPTER FOUR

precisely, was the rationale for focusing on two missed forecasting episodes in order to delve into the social dimensions of meteorological decision making in this chapter. But it is the second episode, where negative example meets exception to the rule, that most clearly exposes sensible weather as a social construction. Let us trace the chain of causation behind the need for a hazardous weather warning in this case. It starts with the momentous traffic impasse *due* to hundreds of car accidents *caused* by the first coating of snow of the season that fell on untreated streets *because* road crews did not arrive in time *given that* the Highway Transportation Department had been caught unawares *because* the NWS forecast called for snow flurries and not snow squalls. In the absence of significant public impact, a one-inch coating of snow does not even meet the minimum criterion for issuing a winter weather advisory for the Neborough region. In light of its public impact, however, a coating of snow constitutes a warnable hazardous weather event. Put differently, by virtue of not issuing a winter weather warning, the Neborough office did not simply fail to predict the snow squalls. More importantly, it failed to predict the chain of events that transformed an otherwise innocuous meteorological phenomenon into a weather hazard. Yet the success of NWS forecasting hinges exactly on foreseeing this kind of chain and preventing it from initializing. Effectively, forecasters are not asked to predict the weather but its aftermath.

A weather forecast, then, is the result of prognostications about social outcomes first and meteorological outcomes second. To be sure, the mutual shaping of the social and the atmospheric typically disappears from view, organizationally distilled already into particular practices and logics of seeing and telling the weather. The institutionalization of exceptional cases, however, where forecasters are explicitly called on to predict the social outcomes of their meteorological predictions, points to the more general lesson: that predicting the sensible weather constitutes such an impossible mandate in practice that, knowingly and unknowingly, the NWS along with the rest of the weather community has sought to mold weather forecasting into a manageable, authoritative task by consistently shielding meteorological decision-making action from the full sweep of its social consequences.

At the Neborough office, Wayne and Phil were already excitedly studying the guidance for the upcoming weekend. This was meant to be a significant snowstorm, successfully handled by the Neborough office as well as the Highway Transportation Department (see Daipha 2007, 170–78). Not that the missed weather event was not still weighing on their minds. But as Phil remarked, comparing what had just transpired with what he was seeing coming toward Neborough on his screens, "those mesoscale things, it was such a narrow band, we are not going to pick up that stuff. That's the bottom line."

The Duality of Error at the NWS

Under conditions of perfect information and perfect certainty, most meteorological decision making would involve a dichotomous choice: to predict or not to predict rain, to warn or not to warn for a tornado, and so on. Alas, information about the weather cannot be perfect, and neither can information about the weather needs of the American public. And so this irreducible uncertainty dulls the dichotomous choice into the qualified but no less pressing question: would it be more beneficial to intervene or not to intervene? Or, in meteorological parlance, "to pull the trigger or not to pull the trigger?" A penchant for aggressive forecasting exposes forecasters to the risk of a false positive, the so-called Type I error. It may well benefit some user groups (such as commercial fishers, as I discuss in chap. 6), but a reputation for false alarms can of course cause irreparable damage to one's credibility. Conversely, the tendency toward conservative predictions exposes forecasters to the risk of a false negative, or Type II error. By definition, a conservative approach offers the best compromise across all forecast users, but an unforecast event, as we have seen, can have dire consequences and even claim human lives. Just as problematically, a consistently conservative approach does not properly engage with the complexities of either the atmosphere or the needs of forecast users and risks being labeled as less than rigorous.

Such considerations, and their implications for their professional legitimacy, have generally led weather forecasters to develop a greater tolerance for Type I error risk because overwarning does not typically result in casualties (Doswell 2004, 1119). So also it appears to be the case for the NWS. Notably, the NWS directives recommend that in "rare situations when the uncertainty of meeting warning criteria is still around 50% within 12 hours of storm onset forecasters should issue a warning or advisory to end the indecision" (NWS 2012d, 3). Ultimately, however, the negotiation of acceptable forecasting risk remains a highly contingent and localized matter because there exist no performance standards by which to arbitrate between Type I and Type II error,[13] neither within the NWS nor the weather community at large. It is instead by raising warning thresholds or amending warning criteria that meteorological organizations have historically attempted to combat over- and underforecasting.

At any given point, therefore, depending on the level of aggregation, the range of decision-making approaches within the NWS generated by the absence of performance standards concatenates into the warning idioculture of a single field office or a single region, a personal forecasting style, and so forth.

Not surprisingly, there are some who lament the loss of forecast calibration and rigor as a result of this state of affairs (cf. Doswell 2004). Yet advocates for the standardization of forecasting risk are, by and large, inattentive to the flexibility and local adaptability that will still be necessary for such a standard to be successfully implemented. As Stefan Timmermans and Marc Berg (2003, 78) have argued for the case of the standardization of medical risk, rather than turning practitioners into "judgmental dopes," standardization would do well to allow them to act *with* the standard: to "act skillfully to match the standard to the actual demands of ongoing work, to keep the standard functioning, and, in doing so, to allow their own work to be transformed through the standard's coordinating activity."[14] Otherwise, the implementation of performance standards amounts to nothing more than a rationalization of practice for its own sake.

Implementation issues aside, the question remains, what would be a defensible ratio of false positives to false negatives at the operating scale of the NWS or, for that matter, the operating scale of a single forecasting office? The issue boils down to what psychologist Kenneth Hammond (1996, 36ff.) calls the "duality of error": at any given degree of accuracy, the number of false negatives can only be reduced by increasing the number of false positives, and vice versa. By implication, regardless of its degree of accuracy, a prediction system characterized by fixed performance standards but irreducible uncertainty is doomed to systematically favor *either* one set of constituents *or* the rest of society. As a result, while forecasters may well be hard at work to improve the overall accuracy of their predictions, establishing the ratio of false negatives to false positives is entirely a matter of policy and cannot possibly be determined on some sort of "objective" meteorological and statistical grounds.

Given, then, the impossibility of charting a course that stays clear of both the Scylla of intervention and the Charybdis of nonintervention, it is hardly surprising that the NWS has been reluctant to institutionalize formal performance standards for managing risk. Yet, although it is not at all clear that such standardization would be worthwhile, it is equally unclear that the absence of performance standards reflects a knowing and intentional policy decision on behalf of the NWS and is not just a product of institutional inertia. For the time being, even as external pressures to

FIVE

Anticipating the Future: Temporal Regimes of Meteorological Decision Making

Ὁ μὲν βίος βραχύς,
ἡ δὲ τέχνη μακρή,
ὁ δὲ καιρὸς ὀξύς,
ἡ δὲ πεῖρα σφαλερή,
ἡ δὲ κρίσις χαλεπή.
ΊΠΠΟΚΡΑΤΗΣ[1]

Throughout this book, I have sought to capture how weather forecasters achieve coherence in the face of deep uncertainty, how they harness diverse information to project themselves into the future. To that end, I have introduced the notion of collage, a heuristic that frames meteorological decision making as a process of assembling, appropriating, superimposing, juxtaposing, and blurring of information. Weather forecasting as the art of collage underscores the culture of disciplined improvisation that characterizes NWS operations. The previous chapter brought weather forecasting to life by compounding the material complexities of the atmosphere with the particularly sensitive social complexities of hazardous weather. And the next chapter delves into the challenges and considerations that go into distilling complexity into a relatable but still authoritative and actionable forecast message. Yet one important question has standardize forecasting performance continue to increase, all organizational effort is directed toward improving the accuracy of NWS predictions and, with it, the odds against overforecasting *and* underforecasting. As for the trade-off dilemma itself, it becomes continuously reinvented and resolved within the micropragmatics of field operations and the decision-making task at hand. With or without performance standards, how could it be otherwise.

yet to be answered: how does the distillation of meteorological complexity vary across different time horizons? If weather forecasting is the art of collage, then which scraps of information get foregrounded when one has to make a decision about the near-term as opposed to the longer-term future? Moreover, how does forecasting practice vary by perceived consequentiality? How, in other words, are projections about the future generated during an emergency as opposed to routine circumstances?

As a first step toward temporally embedding weather forecasting practice, one may identify two dimensions that underlie the logic of meteorological decision making: risk and scale. The first dimension rests on a distinction between "ordinary times" and "extraordinary times," extraordinary times in this context associated with a marked increase in the risk of injury or loss. To be sure, weather forecasters are primed to anticipate and respond to meteorological emergencies. Fine (2007, 20) is entirely on point when he describes NWS operations in the throes of hazardous weather as an "activated organization." Yet, as Wagner-Pacifici (2000, 19) reminds us, an emergency puts stress on any organization, even those organizations established precisely to address it, because "the exact nature, extent, time and shape of the contingent emergencies that do occur cannot be fully anticipated." While episodes of hazardous weather constitute nonroutine but still "normal" events in the context of NWS decision making, therefore, they nonetheless dictate a fundamentally different logic of prospective action due to the elevated threat to human life and property attached to them.

Still, the criterion of risk alone cannot account for the dramatically different approaches to decision making one encounters within both routine and nonroutine forecasting operations. Adding the dimension of spatial scale reveals a pattern to this diversity. The distinct mind-set and course of action typically exhibited by forecasters during summer storms compared with winter storms, for example, can be attributed to the fact that whereas winter storms depend on regional and global atmospheric parameters, summer storms are fundamentally driven by local dynamics— winter storms evolve, summer storms erupt. More broadly, because the more global the reach of a weather phenomenon the earlier its detection, the criterion of scale adds another organizing principle to the temporality of meteorological decision making, regardless of the risk involved.

Transcribed into a two-by-two table (see table 1), the joint influence of risk and scale on weather forecasting practice yields four temporal regimes of decision making: emergency, extended alert, near term, and longer term. By way of illustrating the point that each temporal regime is, on the whole, governed by a distinct logic of decision making, each

Table 1. Temporal Regimes of Decision Making

		RISK	
		Ordinary	Extraordinary
SCALE	Micro	Near-term *short-term forecasting*	Emergency *summer weather forecasting*
	Macro	Longer-term *long-term forecasting*	Extended Alert *winter weather forecasting*

cell has been associated with a particular weather forecasting regimen. In what follows, then, I flesh out and elaborate this rudimentary framework for the temporal organization of decision making in action through its empirical manifestation in summer weather forecasting, winter weather forecasting, long-term forecasting, and short-term forecasting. In the interest of brevity, I will assume the reader is already familiar with the texture of winter weather forecasting (chap. 4) as well as long- and short-term forecasting (chap. 2). I have not yet had an opportunity to discuss severe—that is, summer—weather forecasting practice, however. The next section, therefore, disproportionately the longest of the chapter, also serves as an introduction to the realities of forecasting summer weather for the NWS.

Emergency Decision Making: Summer Weather Forecasting

At the root of what makes summer weather an altogether different animal forecastwise is the prevalence of high levels of solar radiation leading to high levels of heat on the earth's surface. The resulting vertical mixing, or convection, that spontaneously occurs in order to redistribute the heat can cause spectacular levels of atmospheric turbulence the greater the differential between the rising current of warm air and the overlying environment. Because the severity of this type of atmospheric instability is determined by vertical dynamics, the danger it poses is very localized and short lived, making it extremely challenging to forecast, let alone warn for.

The most visible and most violent manifestations of convective turbulence are of course thunderstorms. When the right atmospheric ingredients come together, severe thunderstorms can escalate into hailstorms or they can produce flash flooding, straight-line winds, or tornadoes. Thankfully, the convective weather season at Neborough is relatively mild compared with other U.S. regions both in terms of the intensity and the extent of the severe weather involved. As one Neborough forecaster who had previously worked at an office in the heart of Tornado Alley liked to say, "There is no reason to get stressed out here. Having someone call in to report a one-mile tornado touchdown and seeing nothing on the radar, now, that was stressful." Even at a place like Neborough, however, where tornadoes constitute so-called high-impact–low-frequency events, the explosive atmospheric dynamics, heightened meteorological uncertainty, potentially devastating consequences, and fast-paced decision making of severe weather are responsible for the highest highs and the lowest lows on the operations deck. If there is ever a time when one gets the chance to play god and nail the storm or crash and burn trying, convective weather forecasting is it.

Convective weather, then, requires a different mind-set from NWS forecasters. Time, for one, is assessed differently during the convective season. To say that every second counts is not an exaggeration. Convective weather exquisitely illustrates why the Greeks saw it fit to use the same word, *kairos*, to denote both the weather and the opportune time, the critical "now." *Kairos* captures the sense of things "up in the air" about to descend on us. *Kairos* is the moment of crisis, which, translated from Greek, also means the moment of judgment, of decision making.[2] Switched to convective mode, the radar sweeps fourteen elevation angles in five minutes,[3] and warning lead time is measured in minutes, not hours. No calm before the storm here. Even in the case of mesoscale severe weather phenomena, such as squall lines, which can last for several hours and extend across several thousand square miles, information about which suspicious-looking storm cell will flare up or dissipate often changes drastically from radar scan to radar scan. In the words of a forecaster, "We do not have a fighting chance in the summer. It's hit or miss. In the winter, I pay attention to my timing, and I am always wrong forty-eight hours out. But we do not have a fighting chance in the summer."

The formidable challenge of capturing a severe storm in a timely manner is exacerbated by the fact that space is assessed differently as well during the convective season. As a simple illustration of the problem, consider the distinction between "winter rain" and "summer showers." In meteorological parlance, this involves a distinction between stratiform versus convective precipitation. Rain is produced by stratus clouds, clouds with considerable horizontal extent, and is therefore widespread and continuous, whereas showers are produced by convective clouds, such as cumulus clouds, which are scattered or isolated, hence intermittent or short lived. Per NWS policy, precipitation forecasts indicate the likelihood that at least one hundredth of an inch of liquid precipitation will fall at any point of a given forecast area during the indicated time period. To arrive at such a "point probability of precipitation," forecasters are to multiply their confidence level that a precipitation event will occur by the expected areal coverage of the precipitation. In the case of rain, because of its widespread—that is, 100 percent—areal coverage, a forecast of a 40 percent probability of precipitation indicates that the forecaster is forty percent confident the event will occur. In the case of showers, however, the same probability of precipitation might mean either that the forecaster is 80 percent confident the event will occur across 50 percent of the forecast area or that she is 50 percent confident the event will occur across 80 percent of the area. This is not to say that NWS forecasters actually engage in these sorts of multiplication exercises, of course. In fact, the range and creativity of answers I received when I asked Neborough forecasters for the technical definition of a 40 percent chance of rain showers was rather astounding. The point remains, however, that during the summer season, the factors that go into estimating uncertainty become multiplied indeed, making the decision to warn or not to warn that much more complex.

Here, for once, computer models are of little help. In striking contrast to the winter season, models can at best provide an indication of an increased potential for severe weather. Where and when that may occur is a matter left entirely to the microdynamics of the moment and to forecasters' agility at pattern recognition. The most valuable forecasting tool now is the radar, and forecasters will simultaneously monitor information from multiple radar imageries and sources, obsessively pondering on the structure of an evolving thunderstorm from a variety of angles in 3-D animation, zooming in to check for suspiciously colored regions, then zooming farther in to sample, again and again, the reflectivity value of a storm cell with their mouse as the storm flares up or simply dissipates on their screens.

Because of the atmospheric physics at play, it is only a matter of minutes from the time a convective cell has organized enough to appear suspicious to the time it collapses. As a result, forecasters can, at best, be one step ahead of a summer storm. With every second counting, proficiency at pattern recognition becomes of the utmost importance. Keeping these pattern recognition muscles strong and flexible is therefore the primary goal of the severe weather training meeting conducted early each spring at the Neborough office, in advance of the convective season. Unlike the winter weather training meeting, which focuses on conceptual analysis of model data, the severe weather meeting revolves around visual analysis of observation data. In the 2004 meeting, for example, forecasters were instructed to pay close attention to the shape of the hodograph, that is, the polar coordinate chart used for plotting wind vectors. Subsequently, the "short and stubby," "classic bow echo," and "classic curved" hodographs were identified, indicative of flash flooding, squall lines, and supercells, respectively. In the same vein, radar reflectivity patterns were discussed in terms of a "hook echo" and a "bow echo," while forecasters spent time evaluating the dynamics of "kidney bean" and "comma-shaped" supercells. And they were advised to not hesitate to warn if they saw "a bow echo complex acquire a seahorse signature, regardless of the intensity of the reflectivity."

Meteorological decision making during convective weather, then, disproportionately relies on visual cues from near-real-time observation data, meant to be immediately recallable in their consequences. Indeed, witnessing their training and observing forecasters in severe weather mode, one cannot help but be reminded of marksmen or snipers. At the risk of overstating the analogy, the realities of severe weather promote a "one shot, one kill" mentality. As in sniping, there is no room for second guesses. Forecasters are primed to instantly identify suspicious activity and, if advisable, to "pull the trigger" before their mark has had a chance to do any damage. In practice, the accelerated, compressed temporality of convective weather forecasting tends to strip meteorological decision making down to the basics of trying to keep up with the action. With time running out to make a move, all of forecasters' energy is expended in the effort to decipher the ambiguous information on their screens. No room here to weigh choices and properly make decisions. The urgent, rapidly evolving nature of a convective storm calls for an immediate response, not exhaustive reasoning. Such immediate responses we colloquially attribute to "gut feel," "intuition," or "instinct." This, argues cognitive psychologist Gerd Gigerenzer (2004, 2007), is because they are based on "fast and frugal" rules of thumb: simple and flexible heuristics that rely on minimal information and one-cue decision making to expeditiously tackle uncertainty (see also Klein 1999). Alas, expeditious action invites additional opportunities for errors in judgment, errors that often lead to fatal consequences. The literature is replete with empirical studies of friendly fire accidents, medical mistakes, and firefighting disasters;[4] but errors in meteorological judgment can have fatal consequences as well.

To return to the Hippocratic aphorism that sets the tone for this chapter, the short horizon of opportunity afforded by summer weather makes the challenge of translating professional knowledge into practice especially poignant. In this respect, the training sessions all across NWS offices in anticipation of the severe weather season epitomize an ongoing organizational effort to ensure that the fast and frugal heuristics forecasters will be relying on embody the latest in meteorological science and public service. The objective, always, is to distill and script complex information into cognitively *transparent*—hence recognizable and actionable— diagnostic and prognostic techniques (cf. Gigerenzer, Todd, and ABC Research Group 1999; Gigerenzer and Selten 2002). Yet, within the near-real-time temporal logic of summer weather forecasting, the requirement for fast and frugal decision making aids reaches its purest form. To truly be instantly recognizable, information cues must now be sound or image based, or else they risk not being acted on by forecasters because they will not be processed fast enough by the brain. Whereas textual cues proliferate in other weather forecasting temporal regimes, it is therefore audio alerts and visual imagery that are enlisted, first and foremost, to extract salient informative features out of the noise.

The meterological aesthesis, then, of severe weather forecasting is forged by the radar, which also serves as the basis for the severe weather forecasting collage. Still, information from the radar is regarded as "just guidance," after all. What are desperately needed are reports on the ground, as one thunderstorm may dramatically alter the environment for subsequent storms. Verification concerns aside, therefore, forecasters are eager to know in near-real-time what actually transpired once a storm cell collapses on their screens so as to better assess the dynamics of the situation before their next move. With automated weather stations not equipped to identify hail, let alone funnel clouds and wind damage, citizen volunteers are enlisted and trained by the NWS to serve as severe weather spotters. Once again, agile pattern recognition skills and the proper situational awareness are of the essence. Much like severe weather training sessions for forecasters, severe weather training sessions for spotters aim to translate reporting criteria into easily comprehensible and observable entities. Spotters are thus advised that 50 mph winds "look like small broken branches less than two inches in diameter" whereas 70 mph winds "look like big broken branches, minor roof damage, and whole trees down." Meanwhile, hailstone sizes that meet severe warning criteria run the gamut from a quarter dollar to a hen egg to a tea cup to a softball.

Severe weather spotters are expected to monitor the local NWS Hazardous Weather Outlook and call in to report severe weather conditions as necessary. More often than not, however, it is NWS forecasters who must call after spotters for reports. Because of the highly localized nature of convective events, the spotter network is painfully inadequate for procuring weather reports in remote or low-populated areas, and forecasters routinely find themselves "digging" for more information: calling spotters, police and fire departments, neighboring offices—even family and friends. To be sure, forecasters may solicit ad hoc weather reports during other hazardous weather events as well. But it is during convective weather, when atmospheric dynamics make weather information such a rare, precious commodity to come by that forecasters must—and, indeed, are expected to—chase after ground truth. That is why, in anticipation of a convective episode, NWS offices will typically enlist the help of a local network of ham radio weather chasers, ready to roam the land and bear witness to nature's fury.

Already it should be apparent that the more time-compressed, uncertainty-laden, and dynamic the conditions under which NWS forecasters must decide whether to pull the trigger or not, the greater will be the organizational impetus for collaborative forms of decision making. While the principle of distributed cognition guides all facets of NWS forecasting operations, it truly reaches apotheosis in severe weather forecasting (see Pirtle Tarp 2001). Given the spatiotemporal evanescence of convective weather, the feeling of "controlled chaos" reigning over the operations deck is actually a desired effect, the sign of a team of forecasters at work. Not coincidentally, the emergence of the notion of NWS forecasting as teamwork overlaps with the implementation of a warning team approach to severe weather forecasting, a practice first developed by NWS offices along Tornado Alley. At its simplest, the "severe weather warning team" amounts to nothing more than an impromptu reshuffling of the duties of the staff currently on shift according to the requirements of the developing situation. At its most complex, it can consist of up to ten members and include the storm coordinator, several forecasters on radar duty, a mesoscale analyst, the ham radio coordinator, and a warning disseminator/media liaison.

The rationale behind the severe weather warning team is typically cast in terms of the overall need to efficiently and consistently accomplish a multitude of essential tasks in an extremely short amount of time. There can be no doubt, however, that the overriding emphasis lies in achieving a swift but *comprehensive* appraisal of the situation. In essence, the severe weather warning team serves to multiply the pairs of eyes scanning the incoming radar images for clues on what may be about to transpire. Here the redundant access to weather information characteristic of the NWS operations deck becomes more crucial than ever. Although it is not uncommon for forecasters charged with radar warning duty to be sharing a workstation, when given the option they all prefer to study the radar on their own terms, occasionally wheeling their chairs together to deliberate over a particularly thorny case. In the absence of the temporal continuity that allows hazardous winter weather forecasting to grow in certainty over consecutive shifts, the near-real-time decision making of severe weather must, ideally, rely on concurrent yet quasi-independent evaluations of the situation.

Not surprisingly, the amount of verbal exchange among forecasters reaches a peak during severe weather episodes. Their deliberations, however, do not generate the kind of narrative elaboration of the future, or "foretalk," that David Gibson (2012, 33–34) encountered in his analysis of ExComm meetings during the Cuban missile crisis. Forecasters' talk lacks the if-then expansiveness of scenario building, its projective reach (cf. Mische 2009) thwarted not only by the ambiguous and impenetrable information at their disposal but most importantly by the quick pace of decision making. Staying one step ahead of a summer storm leaves no time for truly exploring alternatives. Pattern recognition and sense making become decision making. The future is now, and once agreement has been reached about what is happening now, the decision about the future has already been made.

Still, despite these extreme time constraints, severe weather forecasting at the NWS is a profoundly deliberative process. Inculcating its forecasters with a particular meteorological aesthesis so that they may be "intuitively" drawn to the organizationally endorsed solution is clearly not enough. Even under such circumstances, leveraging collective insight and effort remains an important organizational goal. The urgency of external deadlines or other forms of "environmental impatience," as Stephen Weiner (1976, 226) reminds us, is creatively enacted by organizations in accordance with internally established norms and procedures. While therefore marked by pressures from the quintessential form of environmental impatience—a summer storm—severe weather forecasting at the NWS remains measured and deliberate, the overconfidence of one's gut feel tempered against securing the consensus of the group. Indeed, deliberation works to keep decision making at a manageable pace—neither allowing forecasters to get carried away in the heat of the moment nor abiding their bouts of indecision.

Yet, the organizational regulation of meteorological aesthesis does not only operate at the level of cognition—notably in the temperance of over/underconfidence bias—but of emotion as well. Gut feelings, after all, are neither pure cognition nor pure emotion but "feeling-thinking processes" (Jasper 2012). And gut feelings are most frequently invoked on the Neborough operations deck during a severe weather episode. Weather forecasters may always reside in the ambiguity and uncertainty of the subjunctive mood, but the extreme impatience of summer storms escalates the "moodiness" (Silver 2011) of forecasting action, exacting the highest emotional and cognitive effort from forecasters. In what is generally considered a stressful work environment, therefore, it becomes especially important that forecasters remain calm and collected during severe weather.[5] Once again, deliberation serves to temper forecasters' emotions into organizationally endorsed action. One's "feel" for the storm will thus be asserted, even exclaimed, on deck, but it will not be acted on until and unless it is recognized as reasonable by the rest of the team.

> Wayne (radar warning duty): I'm gonna go with two county warnings and hope to get one of them. I normally wouldn't do it based on criteria. Norm?
> Norm (radar warning duty): I don't know what you're basing it on.
> Wayne: [The neighboring office] has the criteria, that's the thing that bothers me. My feel is I want to go with it, but I wouldn't go with it based on [vertical integrated liquid]. [*Peter* (short-term desk and storm coordinator) *helpfully points out "a nasty cell" to Wayne*]. I'd sweat it, but we haven't done good with verification recently. I need more reports.
> Peter: But that one cell has a reflectivity of 50–60 dBZ at twenty thousand feet!
> Wayne: Yeah, I know. But it's going to be messy because it's right on the county line. . . . But now the lightning is on! We have positive strikes here! . . . It's a mess, but let's go for it! Let's blast! Thanks, guy! Let's see what happens!

Despite their intense emotional charge, deliberations during severe weather never turn into overt power struggles. Rather, as Fine (2007, 51) has pointed out, NWS standards of collegiality require that authority "be muffled through an *implicit deference structure*." In the above field note excerpt, part of the forecasting episode that will facilitate the discussion for the remainder of this section, note that it is Wayne who is directing the action and not Peter even though it is Peter who, as the short-term forecaster on shift that day, has been formally assigned to the role of storm coordinator. With the seconds ticking away, deference will be interactionally given to the most senior of the group followed by the one considered most versed in the particular weather situation, while flagrant breaches of teamwork decorum will be broached after the fact, if at all.

The ham radio coordinator sits peripherally, his back turned, to forecaster deliberations. He is not part of the decision-making process and is only directly addressed by the storm coordinator, in reality Wayne in this case. Yet, despite his ostensibly marginal position, he enjoys considerable status among Neborough forecasters as he occupies a critical broker role (cf. Burt 1995) within the information exchange network that is Neborough severe weather forecasting operations: he alone controls

CHAPTER FIVE

the flow and content of information between the Neborough office and a ham radio network of weather chasers spanning four states. While he does not overtly participate in forecaster deliberations, furthermore, he may at times prove catalytic for the course of the conversation within the office because his "interactional expertise" (Collins and Evans 2002) lets him anticipate outgoing announcements of reports requested and filter incoming announcements of reports received.

Wayne set the two severe thunderstorm warnings to expire at 3:45 p.m. Norm placed a call over the hotline to inform the Neborough state warning point accordingly, as per procedure. George (ham radio coordinator) transmitted the information over the radio as dictated to him by Wayne, requesting reports of hail and wind damage. It has been eerily quiet for the past forty minutes. The sounds of the ham radio and George's persistent but unrequited call to arms only underline the aggravating static in the room. Everyone is waiting for reports while intently scanning the radar screens for suspicious cells. Wayne says to George, without looking up, "Two miles east of C-town, something is going on. It's a pretty good cell. It looks like something is going on, I don't know what." George dutifully requests reports for hail and wind damage east of C-town over the ham radio waves.

. . .

3:45 p.m. Wayne issues a Severe Weather Statement canceling the two previous warnings. "Nothing happened." At 3:56 p.m., he issues a Special Weather Statement for D-county, the area the storm is moving into. He is waiting for the next radar scan to decide if he is going to issue a warning. Norm says, "There's definite rotation. This thing is a producer."

. . .

Wayne ended up issuing a warning until 4:45 p.m. The neighboring office also issued a warning for the adjacent counties. Peter called them up to see if they had received any reports. Wayne says, "We need info for D-county. Right now. Something big is happening there!"

Just then, George announces a ham report for dime-sized hail in X-town.[6]

Wayne: What? I can't believe that!
Peter: They only have 40 dBZ.
Wayne: I can't believe it.

Bruce (HMT and media liaison) called the nearby airport tower, but they didn't know anything about the hail. Peter is issuing a Special Weather Statement.

Wayne: I don't feel like issuing a warning until I get something better than that.

A second report for dime-sized hail arrives. Now they can actually see the cell flaring up on their radar screens. Norm issues the warning.

Wayne: Alright, any hail reports now verify, go!
George, *speaking with one of the ham spotters*: Okay, it's ending now, but it's still dime sized.
Wayne: Where is he?
George: At C-town, so we can chase this thing east. Quite honestly, he was as surprised as you are [about X-town]. He went and dug these reports from two people.

George is frustrated with the ham guys in X-state [where X-town is located] for not getting the reports sooner: "This is exactly why I'm telling them: 'Dig, dig, dig!' They don't get it. The dynamics of the weather need digging. I hope they get it in the end, or I don't know what I'll do."

Wayne turns around to me: "If we didn't have George, we'd be calling cops all night. . . . My verification numbers won't look good, but I won't sweat it. This is a game—you win some, you lose some. If we don't get reports, we don't get reports." And a little later: "Good reports are helpful no matter what. Even though they don't correspond to what we thought."

Still, more often than not, severe weather reports *do* correspond to forecasters' predictions and warnings. Unlike Fine (2007, 193), I did not witness forecasters asking leading questions when soliciting reports, nor can I imagine such a practice taking place at the Neborough office given its idioculture. Yet it stands to reason that NWS forecasters are going to be quite proactive and resourceful about finding reports to verify their warnings and, conversely, that they are going to resist or even suppress information that does not conform to their hunch that the situation did not merit a warning. In the above fieldwork excerpt, Wayne, Peter, and Norm eventually do "believe" and act on the report of dime-sized hail, but only after it is corroborated by the radar and a second report. This combination of initiative and reluctance when probing ground truth we also encountered in the previous chapter as forecasters endeavored to salvage the verification of the "busted" snowstorm forecast. It is especially pronounced, however, during convective weather forecasting, where the extreme localization of stormy conditions combined with the absence of the systematic, consistent, and calibrated data typically delivered by automated weather stations means forecasters are bound to get the right answers if they are willing to put in the extra effort.

Certainly, forecasters at Neborough appear to have few reservations

CHAPTER FIVE

about gaming the NWS verification effort precisely because they see it as an organizational game to begin with. In this, they readily echo Fine's informants in the Midwest, down to the jokes about "doctoring" the time stamp of received reports or about having George send his men with handsaws out there and then report wind damage back to the office (for a systematic discussion of NWS verification games during severe weather forecasting, see Fine 2007, 178ff.). This ambivalence about the "numbers game" constitutes a leitmotif running throughout NWS forecasters' practice and professional identity, as discussed in chapter 4. Yet it is in the summer season—when the relative ease with which one can game the system, on the one hand, and the elevated potential for tragedy from a missed forecast, on the other, are forced to coexist—that the artificiality of performance measures becomes most exposed and elicits the most conflicted responses from forecasters. When Wayne says above, "This is a game," therefore, he is claiming the license to not only creatively apply the rules but to disregard them altogether if he deems that things are serious and not a game anymore. This is why, although intent on playing it conservatively because "we haven't done good with verification recently," he keeps issuing warnings in the absence of corroborating reports. The spatiotemporal instability and threat of severe weather, the concern for public safety, the excitement on the screens and on the operations deck, all conspire to make the decision to warn or not to warn particularly excruciating. When all else fails, invoking the NWS imperative of getting the message out serves to overcome any last lingering qualms about one's verification performance.

4:45 p.m. The automated weather station in west D-county reports thirty-seven knot sustained winds, and the warning there is about to expire. Wayne thinks the storm is reawakening, but he is going back and forth on whether or not to extend the warning. "I don't know what the hell is going on. We don't get enough reports, so I don't want to overestimate that staff." He calls the neighboring office again to see if they have any verification on the storm. Nothing.

Norm, *to Wayne*: Do you think we're going to verify with sustained winds of thirty-seven?
Wayne: Maybe, maybe not.
Norm: Well, I don't think it's really important if the [automated weather stations] don't pick up on the wind gust. I really don't. For reasons of public safety, I think we should alert people to the possibility. And even if we don't technically verify, we'll still get tree branches, I think.
Wayne: Getting verification will be tough but . . .

Peter, *interjecting*: But you got to let them know it's out there.

Wayne: Yup. Okay, let's do this, but on a conservative approach, and if I see other data we can amend later.

Norm, *muttering to himself*: I really don't think verification is the most important thing.

. . .

6:00 p.m. They just got a ham radio report for lightning strikes. Four injuries. There was no severe weather warning or statement out for that particular thunderstorm as the radar had not picked it up. Norm, frustrated: "That's not good. The [radar scan] angle is no good." Still, a statement had been issued at 1:40 p.m., alerting to the possibility of frequent lightning between 2:00 and 8:00 p.m. So they "are covered," they decide.

But the storm was hardly done yet. Soon after, it flared up again, the screens flashing red with new radar alerts about hail—"Another miss," said Wayne—and it was not until four hours later that the severe weather warning team finally decided to call it a day. Wayne's parting words as he walked out of the operations deck were, "It's been a bad day. We'll see how much damage I did to the office when the numbers are in."

The above field note excerpts of Neborough operations were taken during the severe weather event of August 16, 2003. According to the related one-page Preliminary Operational Assessment produced by the office, "this event was very difficult from a warning perspective as all of the thunderstorms were marginally severe, the primary mechanism for convection initiation being low level instability from the morning insolation." Out of a total of nine severe thunderstorm warnings issued that day, only five verified, based on the bare minimum of severe hail and wind criteria, while three severe weather events that met the criteria were left unwarned. In the next few days, as newspaper clippings were added to the event verification folder, the unwarned severe weather events rose to five.

With the summer of 2003 being the calmest in a decade, this was as bad as it got that season at Neborough. But that arguably makes this episode all the more suited to highlight the character of severe weather forecasting. For the intensity and excitement on the operations deck offer resounding proof that the challenges of severe weather forecasting spring precisely from the *absent-presence* of severe weather. This plain forecasting episode, then, helps bring home the essential properties of meteorological decision making during summer weather. The key ingredient, which sets summer weather forecasting apart from winter weather forecasting, is the overwhelming dearth of reliable information because of the extreme localization of the potentially hazardous event. Forecasters' perceptual field narrows to become highly focused, both spatially and temporally.

This perceptual focus is almost exclusively guided by a single information source, the radar, adjusting and reorienting the action with each fresh batch of data, while "rogue" reports are treated with great suspicion in light of the heightened ambiguity and indeterminacy of the situation. The environmental impatience that results from the extreme localization of the hazardous event to be anticipated, furthermore, militates for multiplied, concurrent yet quasi-independent perceptions—to better probe the atmosphere, but also to better probe the validity of individual judgment. At the NWS, summer weather forecasting strives to provide a reasoned response to an emergency situation.

Extended Alert Decision Making: Winter Weather Forecasting

In important ways, winter weather forecasting constitutes the mirror image of summer weather forecasting along the axis of crisis decision making. Whereas summer weather forecasting must contend with temperamental atmospheric microbursts, winter weather forecasting can rely on a relatively stable, large-scale atmospheric behavior. Whereas forecasters must seize the moment to seize the weather in the summer, in the winter it is trend following and patience that win the day. Summer storms require extra eyes on deck; winter storms are handled by the regular shift crew.

Indeed, winter weather forecasting best fleshes out the relationship between kairological and chronological time. Winter weather still materializes in kairological time, of course. Experiential reality does not exist in chronological time—only abstract, formal entities, such as seasons and climates do. Whereas *chronos* represents a quantitative, orderly sequence, regulated by clocks and calendars, the weather is episodic and situation driven, just like *kairos*. Yet, while seemingly at cross-purposes, *kairos* and *chronos* find themselves intertwined, harnessed together by social actors eager to make time work for them (Bergson 1946; Orlikowski and Yates 2002). When there is little time left on the clock, every second is made to count by transposing and stretching it out onto the matrix of lived experience. And when an event draws out interminably, it gets broken down into smaller, more manageable—albeit arbitrary—temporal units (e.g., Zerubavel 1985). Chronological time may be derivative to *kairos*, which embodies "the fullness of time" (*to pleroma tou chronou*; Galatians 4:4), but, as such, it creates a space for orienting and adapting to the anticipated "fatefulness" (Goffman 1967) of one's decisions.

Winter weather forecasting also fleshes out the relationship between weather and climate. This relationship, to be sure, may quite appropriately be seen as a formal extension of the relationship between *kairos* and *chronos*, but I am more interested here in the *organizational* relationship between weather and climate. Despite their differences in purpose and priorities, which had kept them apart since the nineteenth century, weather and climate information infrastructures have become increasingly consolidated since the 1960s once climate modeling began relying on weather—and not just climate—data for its calculations (Edwards 2010, 16, 289ff.). Today, NWS forecasters at the Climate Prediction Center and NWS forecasters at local forecast offices consult the same global weather models and ensembles, albeit at different levels of granularity, to produce climate statistics and the long-term weather forecast, respectively. And so it comes to pass that winter weather forecasting at the NWS properly begins around mid-October each year, when the Climate Prediction Center issues its Winter Outlook for the upcoming season. Beyond the Winter Outlook, the Climate Prediction Center issues monthly national outlooks for precipitation and temperature events likely to fall outside the climatological average. Its purview, however, is not limited to climatology-based projections. It also encompasses what is conventionally considered to be meteorological territory: issuing national hazardous weather outlooks for days three through fourteen of the forecast. Only then does the Winter Weather Desk of the Hydrometeorological Prediction Center take over to issue national snow forecasts for the remaining days one through three. If the history of the weather can be harnessed to establish the climate, then the climate can be harnessed to establish the future of the weather. With global weather forecast models as their common denominator, where to draw the line between the climate and the weather becomes a fundamentally organizational matter.

Back at the Neborough office, forecasters have been closely monitoring the evolving guidance. Mostly, however, they have been biding their time—despite the mounting anticipation and the objective and subjective weather scenarios proliferating on their computer screens. The greatly expanded horizon of opportunity afforded by so-called long-fused events, of which winter weather serves as the iconic example, translates less into more action than into more *judicious* action. After all, as Schütz (1967, 67–68) has pointed out, the process of decision making entails aiming "rays of attention" onto several probable futures at a time until one scenario frees itself, "like overripe fruit," from the canopy of available options. In the summer, the environmental impatience underlying weather warning operations turns this ripening process into a caricature of itself; so much so, in fact, that a number of scholars have been inclined to substitute the concept of sense making for that of decision making proper (e.g., Klein,

Moon, and Hoffman 2006). But in the winter, environmental impatience turns into environmental patience, and arriving at a decision becomes an exercise in discipline and perseverance. While the first signs of a developing winter storm system are typically detectable weeks in advance, it is still quite possible that this weather scenario will not come to fruition—one had better watch and wait for later model runs before committing to a plan of action. As confidence about the likelihood of an approaching winter storm builds up, furthermore, it is still quite possible that the system will advance faster/slower, that it will produce more/less snow, that it will switch to snow more to the north/south/east/west than currently expected. And so Neborough forecasters hold off making a warning decision as long as possible, taking the time instead to plan and marshal their resources until they recognize an opportunity for action. In the winter, it is not unusual to read in a NWS office's Forecast Discussion, "This continues to look like advisory material tomorrow night, but no issuance at this moment. We have time."

Maintaining a "wise and masterly inactivity" instead of snatching at a decision, however, is easier said than done. The dramatic flip-flopping incident in chapter 4 illustrates that too much time on hand can stretch one's nerves to the breaking point. Purposefully stalling action in an already patient environment only makes the passage of time loom that much larger over the decision-making process. As Wagner-Pacifi (2000, 60ff.) has demonstrated for the case of political standoffs, the exercise of patience while in a subjunctive mood exacerbates the specter of what lies ahead.

To make matters worse, predicting snow introduces a source of indeterminacy into the forecast that cannot be mitigated by the expanded time frame for making a decision. This is because predicting snow also entails predicting the *amount* of snow expected. Even though the passage of time typically brings computer models into better alignment and hence increases forecasters' confidence about the likelihood, track, and intensity of a given snowstorm, there is still the matter of determining how much snow will actually accumulate—that is to say, fall and stick—on the ground. This is a highly consequential decision, indeed the most critical prediction in the snow forecast, for it dictates whether the snowstorm will merit a weather advisory or a weather warning. That NWS forecasters overwhelmingly first settle on what kind of weather alert, if any, to issue before they settle on exactly how much snow to alert for only speaks to the significance, and difficulty, of predicting snow accumulation.

But the most telling sign of the difficulty of predicting snow accumulation is that computer models do not predict snow at all. They only predict liquid precipitation, which forecasters must then somehow convert into its snow equivalent. The oft-quoted rule of thumb for the Neborough region that one inch of rain amounts to thirteen inches of snow is hardly a dependable guide, because this ratio can vary widely—from three inches for wet, dense snow to nearly fifty inches for dry, powdery snow—depending on the slightest temperature change along the atmospheric column. Already, then, per the above rain to snow ratio rule of thumb, missing the forecast by three tenths of an inch of liquid precipitation, something that would normally go unnoticed in the case of a rain forecast, is tantamount to missing the forecast by four inches of snow. And if the forecast is also off by a mere one or two degrees Fahrenheit, the discrepancy in snow accumulation can be quite dramatic. And if the ground temperature is also not as expected, this, again, can lead to a substantial over/underestimation of the amount of snow that will actually stick and accumulate on the ground. Alas, quite often, these highly consequential predictions of a snowstorm's thermal profile can only be determined with some degree of precision within hours of the storm's arrival and do not profit from the expanded horizon of opportunity characteristic of winter weather forecasting. On the contrary, given the iterative character of winter weather forecasting and the profound uncertainty surrounding snow accumulation forecasts, a protracted time frame may overwhelm forecasters into an unceremonious change of heart.

Because winter weather forecasting is therefore a long game, rewarding disciplined prudence and deliberate consistency, the NWS has sought to inculcate these virtuous habits in its forecasters by developing a carefully timed regimen for issuing winter weather alerts. As discussed in chapter 4, NWS offices must wait until forty-eight hours ahead of the storm to issue a winter weather watch, and only if a neighboring office agrees with their prognostic assessment. And they must wait at least another twelve hours before upgrading to an advisory or warning. Ostensibly put in place to promote public preparedness, this staggered warning communication system also serves to structure forecasters' waiting time, making it part of the logic and rhythm of NWS winter weather warning operations (cf. Zerubavel 1981). Kairological time may well expand/contract with more/less time on the clock, but it is the meaningful punctuation of chronological time that allows actors to actually recognize and seize opportunities. By temporally scripting how its forecasters are to enact timeliness and accuracy—the formal criteria for a successful weather warning—the NWS endeavors to mold time into an organizational resource and forecasters into poised decision makers.

This strategy of explicitly marking the tempo of winter weather warning operations so as to slow down the process of decision making brings to mind lessons from dual-process models of cognition, a relatively new paradigm in the psychology of reasoning. Dual-process theory is conventionally explained in formalistic, either/or terms: our brain consists of two separate minds, the one based on an implicit processing system and the other on an explicit processing system. One thus hears of intuitive versus rational, hot versus cold, or fast versus slow decision making. Its expository value notwithstanding, this is of course a false dichotomy, not borne out in practice. What dual-process theorists have, in fact, been claiming is that the implicit processing system underlies all of our reasoning but is at times "supplemented—or sometimes overridden—by the output of the more deliberate, serial and rule-based system" (Gilovich, Griffin, and Kahneman 2002, 16).[7] Still, because intent on exposing the heuristic biases of expert intuition, dual-process theorists tend to overwhelmingly focus on implicit cognition, which they consider in its juxtaposition to explicit cognition. Missing are studies that elaborate the *interaction* between the two systems (but see Leschziner and Green 2013). This represents a crucial gap in the literature, however, as it is widely acknowledged that professional expertise relies on "a combination of analysis and intuition" (Kahneman 2011, 186). The blind spot is partly a matter of method: dual-process models of cognition have been almost exclusively developed through experimental research and therefore lack empirical access into how experts make real-life decisions. Yet a recent opportunity (Kahneman and Klein 2009) to bridge the gap by incorporating ethnographic research on fast and frugal heuristics resulted instead in further alienating dual-process theory from expertise as it actually manifests itself. Drawing a strong line between professions with "real expertise" and professions consisting of "pseudo-experts" (Kahneman 2011, 239ff.) to account for the conceptual disagreement between scholars of fast and frugal heuristics and dual-process theory, respectively, betrays, once again, the optimizing bias of this research tradition in cognitive psychology to the detriment of understanding human reasoning on its own terms. Telling in this regard is the complete absence of any reference to studies of weather forecasters, who, with their skilled predictions in a complex, low-validity environment, would appear to pose a counterfactual to dual-process theory's a priori definition of what constitutes a profession with real expertise.

In fact, the reverse is true. As this book demonstrates, far from a counterfactual, weather forecasting is an excellent case for excavating how analysis and intuition may be combined across disparate temporal regimes to produce skilled predictions. The comparison between summer and winter weather forecasting operations at the NWS is particularly promising in this respect, as it readily complicates the distinction between a fast and implicit versus a slow and explicit reasoning mode. Already, the latest research in dual-process theory challenges previous wisdom by suggesting that reasoning does not have to be slow to be deliberate because decision makers can learn to explicitly adopt fast and frugal heuristics in problem-solving tasks (Evans 2010, chap. 4, 2012). This insight is richly borne out by NWS efforts to instill a proper meteorological aesthesis and decision-making regimen. But there is a bigger lesson to be learned from the NWS case: Whether hot or cold, fast or slow, implicit or explicit, by the time judgments have crystallized into a decision, they have become tempered through particular institutional and organizational expectations and mechanisms of sobriety. While some sociotechnical environments are more effective than others in regulating and supporting decision makers in their task, all are *decisively* involved in shaping human cognition, such as it is, into appropriately externalized enactments of reasoning. The attachment to a distinctly dual model of cognitive processing, therefore, risks mystifying what is empirically, never mind sociologically, salient about the process of decision making. In practice, decisions manifest themselves at the interface between heuristics and deliberation.

Sure enough, NWS practice during ordinary times—namely, during long-term and short-term forecasting operations—offers yet another illustration to that effect.

Longer-Term Decision Making

As common sense would suggest, the further out into the future weather forecasters project themselves, the greater the uncertainty they associate with their projections, ceteris paribus. And yet the long-term desk at the Neborough office is, paradoxically, responsible for the lion's share of the forecast: whereas the short-term desk typically handles the first thirty-six hours, the long-term desk must produce weather predictions for the remaining five plus days. How to explain this grossly uneven distribution of forecasting labor?

The answer is suggested by the law of diminishing returns when combined with the sharp drop in forecasting skill beyond the first thirty-six hours of the weather forecast. As a reminder, a weather forecast shows skill if it is more accurate than an equivalent forecast based on climatology or current weather conditions. Thanks to its massive and long-standing observation infrastructure, the meteorological community finds itself in the rare, enviable position of being able to systematically measure and le-

CHAPTER FIVE

verage advances in forecasting skill. The progressive expansion of the temporal field of weather forecasts—both in terms of range and specificity—constitutes unmistakable evidence to that effect. But so also does the relatively compact future of weather forecasts, which, currently at seven to ten days, is still considerably shorter than conventional long-term decision-making horizons. At seven days, the future horizon of NWS forecasters is close enough to offer them almost immediate, frequent feedback on the outcome of their predictions. Just as importantly, it is close enough to serve as a constant reminder of the foibles and follies of soothsaying. Indeed, more so than the uneven division of labor between the short term and the long term, the starkest illustration of the fact that weather forecasters do not harbor the cognitive illusions of overconfidence typical of most other decision makers is how narrowly they are willing to draw the limit to what constitutes a meteorologically foreseeable future.

The same logic driving the NWS distribution of forecasting effort between the short- and the long-term desk also drives the distribution of effort within the long-term weather prediction task itself. The lack of forecasting skill "this far out" into the future suggests that meteorological uncertainty will be most effectively handled by striving for consistency rather than accuracy. Barring an expectation of hazardous weather during days two and three of the forecast, therefore, the long-term forecaster devotes most of his energy to producing the weather grids for the new day seven. Editing the weather grids for the remaining time periods is kept to a bare minimum—just enough information to provide the broad strokes necessary to establish a meteorological trend. Greater attention to grid quality could only lead to flip-flopping later on, given the lack of skill. In the words of a Neborough forecaster, "The trick is not to get too fancy with it. We are not that good." More detail is progressively filled in as one gets closer to the short-term period, a function of the increased confidence in the predicted pattern that accrues with the test of time as well as the greater number of global and regional models available for the shorter term. But again, the fundamental concern rests with maintaining a consistent "big weather picture."

Effectively, the responsibility of the long-term forecaster is to set up the narrative of the foreseeable weather future. As such, the meteorological challenge facing him is first and foremost a *problem of exposition*: how to best introduce the scene of the weather event(s) to come without overdetermining the outcome. This challenge increases the higher the perceived meteorological eventfulness of the next few days. For, quite frequently, articulating a coherent narrative arc involves avoiding flip-flopping across a series of weather stories as well as within a single one. One must build toward the potential for heavy rain next Sunday without compromising the high winds forecast for the Friday before. The visual aesthetic of the Digital Forecast Database, furthermore, requires that this narrative consistency be negotiated across offices as well.

It is this challenge of narrative consistency that best accounts for NWS forecasters' total and absolute reliance on a single type of model guidance during the four furthest days of the forecast. To be sure, the well-documented plunge in skill beyond day three advises for a highly conservative forecasting approach to begin with, and Neborough forecasters readily admit they "would be lost without a model this far out." Yet the loss in skill alone cannot explain why they are reluctant to algorithmically blend multiple types of model guidance to populate the grids beyond day four—an otherwise common and recommended practice for the rest of the long- and the short-term forecast. What is at issue, rather, is the great number of weather elements/grids that must somehow be cajoled into a coherent narrative plot. Populating all grids from the same model guidance readily recommends itself as the most effective strategy to this end. Hence the undisputed norm at the NWS to rely exclusively on GFS-based Model Output Statistics (GFSX MOS) for days five through seven. At Neborough, forecasters would speak with astonishment of their colleague who was occasionally wont to blend GFSX MOS and GFS raw data for day seven. And back in 2003, when only GFS raw data were available in gridded format for the long term, the decision of one forecaster to exceptionally draw in by hand the Canadian model to populate the long-term grids, as it was significantly outperforming the GFS, was preemptively announced to the entire operations deck and prefaced with the self-conscious qualifier, "I think I'm going to try something stupid here, it doesn't really make sense to get that fancy so far out, but . . ."

As the narrative arc of the forecast unfolds temporally, more detail is incorporated into the forecast: the meteorological complications that sharpen and bring into focus the weather problem at hand. The NWS Directives offer, once again, a reliable glimpse into what are considered standard complicating parameters and when it is considered appropriate to introduce them. Regarding the generation of text forecasts, for example, forecasters are advised that the intensity of precipitation should not be designated beyond day three; that information on wind and visibility is optional beyond the first sixty hours; and that snowfall amounts should only be included for the first thirty-six hours, and only if the probability of precipitation is 60 percent or greater (NWS 2012c).

While the terse and epigrammatic structure of the text forecast discouraged deviation from this canon, the expansiveness of the gridded forecast has invited a more creative application of the rules, with some forecasters and offices regularly introducing additional information into their grids during day four, rather than day three, of the forecast. The Neborough office fell squarely within this category during my fieldwork—consistently more proactive about day four compared with its neighbors. "Day four is very very important for weekend planning" captures the general consensus, and it explains forecasters' frustration whenever a neighboring office would refuse as a matter of principle to entertain amendments to the GSFX MOS grids beyond day three. Even so, the art of forecasting days three and four at the Neborough long-term desk versus the first thirty-six hours at the Neborough short-term desk was still strikingly different. Whereas the short-term forecaster would meticulously work on the forecast one grid at a time, the long-term forecaster would make strategic edits here and there after carefully studying days three through seven, set to run in a continuous animated loop for each weather element. The space for creativity opened up by the new forecasting routine was being used to reaffirm and celebrate, rather than undermine, the distinct logics of narrative consistency and plot development that govern forecasting practice at the long-term desk and the short-term desk, respectively.

Under normal operating conditions, it is safe to assume that the level of acceptable risk a conservative actor, such as a government agency, is prepared to take will be inversely related to the level of perceived environmental uncertainty. Long-term forecasting at the NWS, precisely because of the higher levels of uncertainty and indeterminacy involved, is a tightly controlled and routinized task compared to short-term forecasting. Foregrounded are oversimplified and formulaic scenarios, the processing of so much complex uncertainty cognitively offloaded to computer algorithms embodying the latest in applied meteorology. Tellingly, what forecasters find most challenging about forecasting the long term is the large amount of "drudgework" involved in producing all the necessary grids. The operating principle of "not getting fancy this far out" does not preclude extensive and labor-intensive editing, necessary for making the imported model graphics compatible with the higher grid resolution of the Digital Forecast Database. Long-term forecasting at the NWS is a formidable, time-consuming undertaking. Yet it crucially does not entail the kind of cognitive strain that, according to dual-process theory, is typical of long-duration, uncertainty-ridden tasks. In fact, the Neborough forecasters who said they preferred the long-term over the short-term desk cited the "less stressful" working conditions as the reason behind their preference. Framed as but the first part within the larger arc of the NWS seven-day forecast, predicting the meteorologically distant but foreseeable future has been recast into a fundamentally technical problem, the domain of computers and meteorological informatics, thereby becoming cognitively and organizationally manageable.

Near-Term Decision Making

Meanwhile, over at the short-term desk, the challenge is mastering the quality, not the quantity, of the grids. As the future draws near, the requirement for more, and more specific, information intensifies. The scale shifts from the global/regional to the meso/local, and all energy is directed toward the fine detail of the next thirty-six hours. Progressively foregrounded are reports from the ground, assessed for their representativeness and potential for extrapolation. By now, after the passage of six days and multiple model runs, the broad parameters of the forecast for tomorrow are bound to have fallen into place. Yet sometimes, for particular weather systems especially, serious uncertainty can persist into the eleventh hour, and the latest surface observations combined with the continuing divergence among the models can lead to an overturn of the status quo in the hands of an aggressive forecaster. Be that as it may, with the narrative arc of the forecast approaching its climax, expository coherence becomes trumped by plot development. The growing anticipation of what is to come escalates the need for more meteorological detail, but the additional information is not simply filled in—there are no preexisting blank spaces in the forecast narrative or the Digital Forecast Database. Rather, consistent with the heuristic of collage, "details" *transform* the forecast into a new gestalt, compelling forecasters to commit to the emergent weather scenario simply by virtue of weaving together the latest information. Such transformations may be subtle or profound, yet what are subtle changes for one set of NWS forecast users can of course have profound implications for another. Even forecasters who resist composing their forecast with a particular audience in mind are keenly, if vaguely, aware of this fact. The general consensus is that it is these last twenty-four to thirty-six hours that make or break the forecast. "The forecast is in the details," as one Neborough forecaster put it.

A corollary of this consensus is that the forecast is forecasters' to make or to break. If their contribution at the beginning of the forecasting task looks dangerously similar to clerical work, the last thirty-six hours gloriously restore NWS forecasters' claim to professional expertise and authority. Computer models, up to this point driving the decision-making process, are now bound to remain locked in a trend and offer little in the way

CHAPTER FIVE

of new information. On what is hopefully by now a robust baseline, it becomes incumbent on forecasters to elaborate on the specifics of the forecast. To do so, they will enlist their knowledge of meteorology, their skill at pattern recognition, their experience with the meteorological peculiarities of the region, and additional local observation data. In principle, this is what they have been doing all along, of course. The forecast is just as much enhanced by humans as it is by machines. At the NWS, certainly, forecasters augment the capacities of computers in the same measure as computers augment the capacities of forecasters, the boundaries between them exquisitely blurred in practice, so that distinguishing between the two becomes an artificial and rather pointless exercise. Nevertheless, the reorganization of NWS forecasting labor along the temporal axis[8] and the distinct forecasting logic of each workstation serve to maintain and reinforce the long-standing comparison between forecasters and computer models.

By extension, the long-standing distinction between "good forecasters," who treat computer models as just guidance, and "safe forecasters," who treat computer models as the final word, has now been boiled down to one's performance at short-term forecasting.

You have to have a framework of how to approach the day, starting with the big picture, understanding what's driving the atmosphere this week, where are we at. Be able to take that into where the sizeable weather is happening. Some people are very, very good at that. Others unfortunately have been slaves to the model. . . . And they're going to mimic what the model gives them, and you wind up with a very safe forecast with very little value added. The good forecasters, the excellent forecasters, they don't stop there. They start higher in the atmosphere. They look at the global pattern to begin to assimilate what's driving the bus. Where are we? And bring yourself down then into the lower atmosphere and begin to assimilate your own model within your head of where things are. And you can use the ensemble models to help you couch this. You begin to sense that, okay, we're in this type of regime, we're in a transition. Transitions usually bring excitement. What kind of excitement are we in for? You come down into the models then and you start looking at the fronts and jets, the vorticity advection, and you begin to look and view how the model graphics are simulating what the atmosphere is going to do, and you begin to see where your weather makers are going to be, and you come down and bring that down to the surface to where we're really forecasting, the human impact, the impact on the surface of the planet. And you're assimilating tons of information. . . . And then you finally get into a point where conceptually you have an idea of what you're going to forecast. Communicate that to your [neighboring offices], and then you go into the Graphical Forecast Editor, and you assimilate those pieces of information to best fit what you're going to forecast. . . . Others, others, the ones who don't have that innate ability, struggle, especially in times of transition in patterns. They're going to go on old faithful: surface thicknesses and precip[itation]. And they're going to play it safe, and a fair percentage of the time they won't be that bad. But they won't be adding that value. They won't be making that big call that could have a significant impact.

Tellingly, Neborough forecasters who thrive in the hot seat of the short-term desk enjoy higher prestige among their colleagues. In fact, those forecasters who expressed a preference for the short-term over the long-term desk were also consistently singled out as "a good forecaster" by the forecasting staff.[9] The relationship between professional reputation and workstation preference is mediated by perceptions of forecasting performance, of course. Yet the point is that such perceptions are critically shaped by the ecology of forecasting practice itself. The reorganization of forecasting labor has meant that the "fatefulness" of meteorological decisions is now overwhelmingly concentrated at one workstation. To continue borrowing from Goffman (1967, 217), because the short-term desk is where the action is, it also monopolizes opportunities to test and display "character": namely, to showcase one's forecasting skill and "maintain full self-control when the chips are down." The temporal splitting of the NWS forecasting task thus segregates and clarifies expertise, hence making it more visible.

This, I must emphasize, is not to say that the temporal splitting of NWS forecasting labor maps onto a split between drudgework and "edgework" (Lyng 1990). One should be careful not to conflate forecasters who thrive in the hot seat of the short-term desk with voluntary risk-takers. Rather, effort is where the action is. I am still following Goffman's definition of action here but explicitly limiting its scope to *cognitively* straining "problematic" activities. If one agrees with Goffman (1967, 185) that action involves "activities that are consequential, problematic, and undertaken for what is felt to be their own sake," then effort is a necessary condition for the experience of action and, therefore, for a feeling of self-determination (Goffman 1967, 214). This reading better grounds Goffman in the pragmatist tradition by still linking agency with effort (cf. Emirbayer and Mische 1998) but by also acknowledging that the vast majority of engaging, even thrilling, activities do not entail cognitive strain (Silver 2011). Removing thrill seeking from the forefront of action, furthermore, rescues Goffman's conceptualization of high-risk professions from an unfounded valorization of risk (cf. Desmond 2007). No less importantly for the purposes of a study on decision making, this reading makes Goffman's action

CHAPTER FIVE

theory compatible with the "fast and frugal" heuristics and dual-process traditions in psychology.

Forecasters who thrive in the hot seat of the short-term desk are consistently more willing to push the limits of their comfort zone, cognitively tackle the vicissitudes of an inherently uncertain phenomenon, and "mak[e] that big call that could have a significant impact." That is why they are considered better forecasters by their peers. In an organizational culture that does not reward specialization, insisting instead that every forecaster be prepared to anticipate the future across multiple horizons of decision making, the simple act of choosing the short-term over the long-term desk pointedly encapsulates one's commitment to mastering the inevitable challenges ahead. And it represents the clearest of signals that, given enough time, this person is bound to actually be more skilled at making predictions, and not just more skilled at handling the pressures of the job.

SIX

Whose Weather Is It Anyway? From the Production to the Consumption of Decisions

I like to think that I have a weather eye. My passion, as a recreation, is fishing, and all fishers spend a great deal of time considering the weather, trying to work out what the weather is doing, suffering the weather and abusing the weather. We sniff the air, study the track and speed of the clouds. If the wind is in the north or the east, we frown. If the sun is blazing from a cloudless sky, we frown. If rain threatens, we frown, and if it arrives, we curse. We are very rarely happy about the weather and very often unhappy. But I would not say that our unhappiness runs very deep, for it is most unusual for the weather to be so bad that it actually stops us fishing.

TOM FORT, *UNDER THE WEATHER* (2006, 8)

Unlike several other national weather services around the globe, the NWS of the United States is funded entirely by taxpayers' money. Its services belong directly to the American public, truly making it—as the slogan would have it—"The People's Weather Service." Certainly, the NWS is ostensibly organized around meeting the weather needs of the nation. And its performance evaluation is not only based on the accuracy of its forecasts and warnings but on its communication impact as well. Indeed, as already discussed in chapter 1, the professional identity of NWS forecasters appears to be fueled as much by a commitment to public service as by a commitment to meteorological science. To be sure, it is the private weather industry that, through its presence and advertising in the media, holds the public's imagination on matters meteorological today. The "mediatization" (Weingart 1998) of the weather forecasting enterprise, already in full swing in the early seventies when the NWS ceased directly broadcasting its forecasts over commercial television and radio,[1] has driven a deep wedge between the American public and the NWS message. And yet, as frustrating as this state of affairs may be for NWS forecasters, it pales in comparison to what, they will tell you, is the bane of their existence: public complacency over weather warnings and predictions.

Evidence to suggest that there is indeed cause for concern abounds. According to the results of the 2012 Federal Signal Public Safety Survey, "when it comes to taking action, despite receiving a notification, just under one half (47%) of Americans would be motivated to take action during a warning of potential severe weather," "more than one in four respondents (28%) would require confirmation of severe weather, such as an actual tornado sighting, flood waters or a visible fire in order to take immediate action," while one in twelve respondents said "nothing would cause them to care" (Federal Signal 2012, 8). The persistence of jokes about the unreliability of weather forecasts, conclusive evidence to the contrary notwithstanding, and the inconsistent response to NWS advisories and warnings are further testaments to the fact that successfully tackling the uncertainty of the weather is not enough for producing a good forecast. The NWS must make its pronouncements meaningfully relevant to its various publics in order to capture their attention and, not least, in order to justify its existence as a federal agency.

Initiatives to redress the situation have been overwhelmingly based on what one might call a "public mistrust" approach. They have been overwhelmingly focused, then, on educating the public on how to recognize and prepare for potential weather hazards and on improving the accuracy of NWS predictions. Meanwhile, in one Pew survey after another, about 50 percent of respondents say they follow weather news "very closely," with the next most popular news category trailing behind by about 20 percentage points (Pew Research Center 2012, 30). Weather news is big news, hence big business. The American public cannot get enough information about the weather, yet it tends not to heed the advice of the experts dispensing it.

This is a false paradox, however, and it is worthwhile to jump ahead for a moment to say that my interviews with commercial fishers in the Neborough region will provide evidence *against* a model of weather risk communication that is premised on the assumption that the public mistrusts science. Echoing recent scholarship on hazards mitigation and disaster preparedness, I will argue instead for the necessity to communicate uncertainty to the public in order to preserve and, indeed, promote credibility, legitimacy, and trust. Yet, even as I will demonstrate that the NWS solution to gaining the confidence of a restive public is misguided, I will also be demonstrating that—to put it somewhat provocatively—this manner of performing meteorological expertise in the public domain is nonetheless bound to persist.

This chapter, then, rounds off the discussion on meteorological decision making at the NWS by turning to consider the production of the NWS forecast within the context of its consumption. In the process, it inevitably challenges conventional distinctions between "expert" and "lay" decision making, proving once again that skilled problem solving is intimately context dependent. But the major thrust of the chapter still revolves around the challenge of creating an expert prediction. Indeed, at no time is this challenge more concrete—or the logic of meteorological decision making at the NWS more transparent—than when examining what makes for practically useful, therefore actionable, expert meteorological advice.

The NWS and Its Publics

The NWS finds itself in a formidable predicament. Based on its mandate, it must protect the life and property of the entire United States. To be sure, with its decentralized structure of 122 forecast offices spanning the land, the NWS appears well positioned to attend to its widely diverse sets of audiences. Especially since its operational guidelines recognize meteorological risk as highly context dependent—not only in spatial but also in social terms. One thus reads in the NWS Directives, "A small craft advisory may be issued for sea/wave conditions deemed *locally significant*, based on *user needs*, as long as the waves threshold is not lower than 8 feet" (NWS 2012c, 14, emphasis added). And the NWS website warns mariners that "There is no precise definition of a small craft. Any vessel that may be adversely affected by Small Craft Advisory criteria should be considered a small craft. Other considerations include the experience of the vessel operator, and the type, overall size, and sea worthiness of the vessel" (NWS 2012a). Correspondingly, at the regional level, the lowest threshold for issuing a small craft advisory ranges from five feet of seas/waves in the eastern United States all the way to ten feet in Hawaii.

Even when the United States is broken down into 122 pieces, however, the challenge facing each NWS forecast office is no less formidable.

CHAPTER SIX

The "public" is still composed of infinite crisscrossing publics, publics that seem doomed to remain largely amorphous and elusive. The fact that it constitutes a public-domain science ironically would seem to structurally preclude NWS forecasting from any extensive direct communication with its publics, dictating instead a fundamentally unilateral flow of information. NWS calculations of meteorological risk thus tend to overwhelmingly rely on functional constructs of NWS forecast users, or "imagined lay persons" (Maranta et al. 2003), as a way to counteract the lack of actual information about the characteristics and needs of the American public. As such, even though quite often the result of extensive and belabored deliberations, conceptualizations of forecast users are rather flat and overly simplistic.

This state of affairs is, naturally, a source of some unease among NWS administrators and forecasters, most often vocalized during my fieldwork in the course of discussing the transition to the gridded forecast. As one high-ranking official confided,

What probably we do not do such a good job of, what we could do a better job of, is somehow being able to extract from our customer, have them be able to project and brainstorm and be creative about "If you can have whatever you want, what would you like?" There's a danger, theoretically, that the [Digital Forecast Database] would be a mistake, theoretically. That we'd go off and run afield and make ourselves irrelevant.

Yet it would be misleading to assign the scarcity of information reaching the NWS about the real weather requirements of the American public to a structural explanation alone. Structural constraints to information flow cannot fully account for the remarkably vague and evasive answers I received whenever I would press Neborough forecasters on the demographic characteristics, never mind the actual needs, of the publics they were trying to serve. And the Neborough case is not an isolated phenomenon. According to a recent NWS survey, more than half of NWS forecasting offices are not familiar with the census data for the populations in their jurisdiction (Fine 2007, 275n8).

That the NWS has thus far paid little systematic attention to the forecast use and needs of the American public must, in part, be seen as a reflection of the chronic neglect of the social dimensions of weather forecasting in the meteorological community at large (Pielke and Carbone 2002; Hooke and Pielke 2000; Katz and Murphy 1997, preface). But just as turning the issue into a structural communication problem does little to account for NWS forecasters' reluctance—if not downright resistance—to educate themselves about their audiences, the invocation of professional culture and organizational history as additional forces structuring the logic of NWS forecasting practice cannot quite account for the persistence of this reluctance in the face of mounting external pressures.

Rather, to understand how NWS forecasters can be at once almost willfully ignorant about their publics yet genuinely committed to public service, one must reframe the issue as a *problem of expertise*. Maranta et al. (2003, 151) succinctly summarize the problem—or, more precisely, the dilemma—as follows: experts must maintain an epistemic asymmetry between lay persons and themselves in order to appear credible; but they must bridge that epistemic asymmetry if they are to provide useful advice; the only way to successfully bridge the asymmetry, however, is for them to appear credible.

While this dilemma confronts all experts, it is particularly acute for NWS forecasters and other experts in the public domain. By definition, the process of NWS forecasting cannot be disassociated from the publics it is meant to serve. The NWS requires the loyalty and cooperation of the American public, or it has no reason for being. As befits the People's Weather Service, therefore, the meteorological advice dispensed by NWS forecasters comes bundled with all sorts of explicit and implicit social assumptions and expectations so that it may become accessible and relevant, hence *relatable*. Yet it will not do for NWS calculations of meteorological risk to simply be accessible and relevant to their intended audiences—they need to come across as *actionable*, too. And for that to occur, NWS forecasters much fashion their advice such that it be perceived as competent and legitimate. They must, in other words, establish enough of an epistemic asymmetry between themselves and their audience that their pronouncements will command respect, hence become actionable. Rephrased in the terms of the NWS case, the problem of expertise is: How to relate actionable meteorological information? How to distill and predict atmospheric complexity in terms that are conversational yet still authoritative?

Brought down to the level of practice, the problem of expertise for a public-domain science such as NWS forecasting turns the production of imagined lay persons into an indispensable part of the day-to-day production of the weather forecast. More than a pragmatic device for dealing with a multitude of widely diverse audiences, constructs of the "public" allow NWS forecasters to maintain a safe social distance that reinforces their privileged epistemic status and normalizes their power to designate "common sense" responses to their predictions and warnings. Meanwhile, in the increasingly competitive marketplace of weather forecast providers, the push for innovation—coupled with a professional culture that remains, by and large, oblivious to the social underpinnings of its services—further entrenches this tendency toward an "expert knows best" approach and a reliance on functional constructs of forecast users.

NWS forecasters' disposition to rely on functional constructs of their publics raises the inevitable question: how well do such constructs stand up to reality? That is the question that motivates this chapter. Taking it up requires a foray into the lifeworld of forecast users, something I will pursue in the second half of the chapter. But first it requires a look into situations in which forecasters' constructs of the weather information needs of their publics are actually confronted with the reality of these needs. For, far from intractable and monolithic, such constructs must be continuously adapted and realigned to practical considerations. Situations that challenge and complicate forecasters' assumptions about their publics, therefore, allow insight into their logic and scope. And such situations are many. Exactly because it is the People's Weather Service, the NWS must heed, or at least it must appear to heed, the voice of these very same elusive publics.

NWS Publics by the Numbers

The formal evidence of that voice is predictably elicited through surveys. Traditionally, it fell on local forecast offices to conduct NWS surveys.[2] They were overwhelmingly based on convenience sampling techniques and did not aspire to statistically generalizable results (Van Bussum 1999). This was no doubt precipitated by a lack of adequate manpower and resources, but it is clear that the real objective was to solicit feedback of any kind, the underlying assumption being that NWS audiences are too vast and varied to be properly studied.

The Government Paperwork Reduction Act of 1995 effectively put an end to local survey initiatives. NWS surveys are now conducted exclusively online, designed at the regional or national level, and hosted on local office websites. These surveys are equally problematic, however. In the spring of 2004, for example, the Neborough office ran a "NWS Web Site Customer Satisfaction Survey," a twelve-item questionnaire soliciting feedback on the appearance, organization, information accessibility, and quality of the Neborough website. Given that 327 out of a total of 351 respondents reported visiting the website on a daily basis, it is hardly surprising that the overwhelming majority of answers were evenly split between "satisfied" and "extremely satisfied." This was the only survey conducted during my twenty-two months at Neborough, but it is fair to expect that, at a minimum, all such online survey efforts will be suffering from a "sampling on the dependent variable" bias.

Not that there is an absence of more rigorous and systematic surveys on the publics of the NWS. In compliance with the 1993 Executive Order 12862, "Setting Customer Service Standards," the NWS has employed the services of a management consulting firm to obtain American Customer Satisfaction Index (ACSI) measures for segments of its audience.[3] Emergency managers, media, aviation, and marine customers were surveyed for the first time in 2003, followed by the general public and climate and fire weather customers in 2005. NWS scores have generally been well above the aggregated federal ACSI score as well as the overall ACSI score, with the general public score hovering at about eighty-five points and the marine score at about eighty points out of a hundred.

The majority of NWS forecasters are not aware of these results, however. National surveys stay beyond their radar, even though survey reports are circulated electronically among forecast offices. At Neborough at least, most forecasters said they "sort of assume they exist but don't know for a fact," while the few forecasters who were aware of their existence regarded them as mere popularity polls, of little substantive value to their daily forecasting concerns. Questions about their validity aside, the results of national surveys appear to be too diffuse and abstract to be of immediate relevance. With local survey initiatives not an option anymore, it is through direct interactions that Neborough forecasters give a face to their publics.

"Interacting with Customers"

The weather is everyone's business, and Neborough forecasters are literally surrounded by opinions on the matter. One consequence of this state of affairs is that it has forced forecasters to quickly grow a "thick skin," as they like to stoically admit, in order to field the obligatory wry remark over the latest missed forecast. Another consequence of this state of affairs is that it tends to lull forecasters into a false sense of security about their knowledge of their audience:

Phaedra: Do you know what NWS customers need and want?
Forecaster: I think I interact with them a fair amount.
Phaedra: Really? How?
Forecaster: When I go out, like in a bar. Say, conversations with a girl. It's a pain, I
 don't bring it up, but they're like, "What happened to the forecast, you had show-

ers and thunderstorms." Or at the [baseball] game, several of my friends were asking me about it, and I also heard two independent conversations. So I know. That's getting feedback. That's how I knew that our forecast of "showers and thunderstorms likely" did not really do that good of a job.

Several weeks into my fieldwork, I began to notice that in the course of talking about "NWS customers," Neborough forecasters would in fact differentiate among three distinct audiences: "the general public," "partners," and "customers." As the weeks progressed, this emic classification scheme would prove critical for tracing how forecasters construe and negotiate their social responsibility as protectors of the nation on a day-to-day basis.

Not unlike the rest of us, Neborough forecasters invoke "the general public" as a convenient blanket term for the multitudes living and playing in the area of their forecast responsibility. When not defined negatively, the notion of "the general public" is in effect populated by the characters that regularly inhabit forecasters' lifeworld. The family members, friends, and acquaintances, always happy to volunteer their two cents. The fifteen citizen weather observers and the pilots of five ferry lines who, in the course of reporting their daily observations, will often comment on today's or yesterday's forecast. And one should not forget the phone calls from those whom Neborough forecasters refer to, sometimes affectionately and sometimes resentfully, as "the regulars": "the X-town lady," the eighty-year old woman who calls in almost daily seeking reassurance at the mention of a slight chance of hazardous weather in the forecast, or "the bookie from Metrocity" who requires detailed information about the weather conditions at kickoff and the exact timing of the precipitation forecast. Taken together, all these characters reinforce forecasters' impressionistic opinion that lay notions of what is a good forecast are based on a superficial and simplistic understanding of weather and weather forecasting and that the general public is better off being told in categorical yes/no formulations what the weather is going to be and be spared of all the uncertainty.

Interactions with the general public during outreach events only serve to further confirm this opinion since their primary objective, in the words of a Neborough forecaster, is "to spread the gospel out"—to educate about weather hazards and advertise available NWS services. I observed over a dozen such outreach events, ranging from a boat show exhibit and a StormReady Community certification to high school talks and weather spotter training sessions, over the course of my fieldwork. Forecasters would occasionally get to hear from the audience, but these occasions were not structured into the event, and therefore feedback was hardly systematic. Despite Neborough forecasters' sincere professions that they are interested in what the public has to say, the entire experience is structurally hierarchical and monologic.[4] At best, the audience may be seen as sharing a fascination about the weather with forecasters, but theirs is implicitly understood to be a naive and amateurish fascination. Their presence alone is taken to indicate that they came to be enlightened by the experts on the subject.

It would be fair to conclude, then, that Neborough forecasters' interactions with the general public are based on what has come to be known as the "deficit model" of the public understanding of science, which—because it assumes "public deficiency, but scientific sufficiency" (Gross 1994, 12)—casts the public in the role of passive and rather ignorant consumers of scientific information (Wynne 1991; Ziman 1991). Although originally premised on a trusting public, the deficit model is in fact only bolstered by a supposed public mistrust in science since this mistrust is taken to stem from public ignorance and misunderstanding in the first place (Jenkins Report 2000). Because it naturalizes an epistemic division of labor between experts and lay persons, between those who know the weather and those who are told about the weather, this point of view grants forecasters broad authority to determine which simplifications are "appropriate" for getting their message across (cf. Hilgartner 1990, 520). It thus readily recommends itself as an expedient solution to the problem of expertise plaguing the NWS and other public-domain science agencies.

Yet, in practice, not all NWS publics are treated as equally "deficient." If the boundaries between experts and lay persons stay firmly in place during forecasters' interactions with most consumers of their forecast, they quickly dissolve during interactions with "NWS partners"—a term NWS forecasters use to refer to other government agencies, the media sector, and the weather community at large. The setting now is a round table instead of a podium, and the primary objective becomes to cultivate and nourish an atmosphere of open communication and partnership. Here, any lingering concerns about asymmetries in expertise become renegotiated around the NWS motto of "Working together to save lives."

I encountered perhaps the most striking example of these dynamics during my very first meeting between Neborough forecasters and local television meteorologists barely a week into my fieldwork and with reports of the uneasy relationship between the NWS and the private weather sector weighing on my mind. A considerable portion of the meeting was still taken up by Neborough forecasters advertising and educating the media about the benefits of upcoming NWS services, notably the new hazardous

CHAPTER SIX

weather alerting system and, of course, the Digital Forecast Database. But this meeting also served as a communal brainstorming session on how best to preserve the quality of the long-standing Neborough climatological record in light of recent problematic snow measurements at Metrocity International Airport. And further, it was unanimously decided that, "for the common good," television stations would share the data from their radar with the Neborough office. More importantly for the purposes of this discussion, media meteorologists were expressly, and repeatedly, asked about their requirements from and complaints with the Neborough office. In fact, when the request for more frequent snowfall amount updates was met with the response from a Neborough forecaster that NWS surveys have shown the media are happy with three- to six-hour updates, an uproar ensued: "Why do we care what the Weather Channel wants? They are in Atlanta! When there's even half an inch on the ground, my producer is pumping us for snow totals every hour, nothing else. I'd rather have overload than underload!" In the end, it was mutually agreed that the Neborough office would strive to issue snow amount updates at least once every two hours, an agreement the Neborough office has abided by ever since. Meanwhile, there was much bonding over the frustration at being "forced into a seven-day forecast," a threat to one's professional credibility made worse by the "low weather IQ of the public."

This epistemic symmetry is not exclusive to forecaster interactions with "NWS partners," however. It also resurfaces during interactions with a certain class of end users. In the course of my fieldwork, I witnessed several meetings with what has increasingly come to be associated with the term "NWS customers"—big corporate entities, such as gas and electric companies. One such occasion was the visit by Rick Curtis, the flight dispatch manager at Southwest Airlines,[5] who holds a bachelor's degree in meteorology and membership in several professional meteorological associations. His visit to the Neborough office was partially motivated by a recent incident where, because of a Neborough forecast of low visibility at B-town airport that stranded Southwest planes in A-town, three different Southwest dispatchers called in to complain, trying to convince John, the Neborough forecaster on aviation duty that day, to change his forecast. John was still fuming over the incident. When he heard that Curtis had phoned to apologize and intended to come to the office to discuss the situation, his reaction was, "The meeting will be useless because what will end up happening is he will come in to preach his case rather than listen to us. The bottom line is they are concerned about profit first and safety second, while we're concerned about safety. Because if a plane goes down, our butt's on the line."

At the exceptionally well-attended meeting, Curtis started off by reassuring the staff that "We want a partnership with you. We want to coexist and thrive." For three hours, he painstakingly went over Southwest's flow of activity, paying particular attention to how NWS information is used, so that Neborough forecasters would get a sense of "what you guys are up against." Inevitably, the conversation turned to the contentious subject of TEMPOs—aviation forecast segments meant to indicate a temporary fluctuation in weather conditions. NWS Directives remind forecasters that "the lowest meteorological condition contained in [an aviation forecast], regardless of any conditional language, will drive operational decisions" (NWS 2005c). Forecasters are, therefore, instructed to use TEMPOs sparingly and for high probability expectations only. Curtis said he looks at a variety of NWS products, the most important being the Area Forecast Discussion because "I can almost get a confidence value out of that because you present alternatives." And he continued:

What we hate are TEMPOs because it tells you nothing. In a perfect world, I would eliminate TEMPOs. Give us your best shot, and, if it's wrong, issue an amendment. Because any time there's a TEMPO going in there, we have to carry extra fuel for an alternate [airport]. For the private pilot, TEMPOs are a big deal, but not for us. And everybody has to use the same [aviation forecast]. Now, I realize this is the most controversial part [*here Curtis took a dramatic pause*] but I want you guys to forecast weather the best you can and try to get safety out of the equation because, from our perspective, we're covered. The safety should be up to the user of the information. Your job is to give us the best, most accurate forecast you can. By policy, we cannot but pay attention to safety, we just can't do it. We've never had an accident. Safety is huge, that's what a carrier's reputation is based on. It's more important than customer service. We wouldn't be around for thirty-two years if we didn't care about safety.[6]

"Our philosophy is to be the most accurate we can," responded the director of the Neborough office, amid nodding heads. "Otherwise, it's going to hurt our verification and credibility. It's just that we sometimes think we might not have the best information we'd be comfortable with, and so we think that for a brief time things might be bad. We just don't have a better way to communicate that fact."

At the end of the meeting, everyone agreed this had been a productive discussion, and mutual pledges were made to "keep in mind what we learned today." Whether Neborough forecasters and the Southwest representative reached a substantive agreement or not was not of immediate interest. The true objective for all concerned, after all, was to keep the lines of communication open. The important achievement was that both parties were able to voice their concerns, express their needs, and mend broken fences. Here, at last, one begins to recognize what Nowotny et al. (Nowotny, Scott, and Gibbons 2001) contend is the emergence of reconfigured forums of socially distributed expertise, where scientific data and the realities of multiple publics are juxtaposed to generate socially robust knowledge.

Perhaps the biggest challenge facing NWS forecasters as public-domain scientists is that their publics are too numerous to be accommodated within a single calculus of risk. Yet the public arena of NWS forecasting is hardly as open and participatory as one might expect, or require, of a public-domain science. Rather, it emerges as insular and hierarchical, dominated by the interests of political and industrial stakeholders. Although it is the needs of the general public that are still foregrounded symbolically and rhetorically, it is the needs of customers and partners that are actually *listened to* in practice. As Fine (2007, 211) already noted several years ago, the NWS increasingly looks like "an agency that operates as an adjunct to industry." Thus, the question looms large, and it is a question that has been looming in the background since the first pages of this book: How well correlated is NWS improvement in mastering meteorological uncertainty with NWS effectiveness in protecting life and property?

To put the issue in terms of the problem of expertise facing NWS forecasters: How do their functional constructs of the public and its weather needs fare beyond the institutional walls of the NWS? How robust are they in the face of lay assessments of the value of the NWS forecast? To begin to answer these questions, I will now venture beyond the citadel of the NWS and the Neborough office and turn to the point of view of commercial fishers working in the Neborough region. Mariners are arguably the last remaining group of forecast users directly dependent on the NWS for weather information because, by all accounts, it is only the NWS that issues a marine forecast proper. Attending to the experiences of commercial fishers offers therefore a unique opportunity to move beyond functional constructs of NWS "publics" and concretely gauge the analytic triangle of forecasters–weather–forecast users.

Commercial Fishers and the Weather

Neborough forecasters readily agree that mariners' sophisticated use of weather information sets them apart from most other audiences in the general public. Indeed, the sheer number of weather adages and other lore surrounding the seafaring world attests to the perilously intimate relationship with the weather elements mariners have to contend with and the forecasting skills they have been forced to develop.[7] Commercial fishers especially are renowned for their weather savvy—almost as much as they are renowned for their audacious battles against the sea. If commercial fishing is founded on taking on the daunting challenge of somehow carving a living from the sea, then learning how to deal with the weather is an inextricable part of the practice and the culture of being a commercial fisher.

Fisher: Years ago I didn't fish as hard. Now, I fish in more weather, you know what I mean? When I first started fishing, you know, I used to turn around a lot more.
Phaedra: And what has changed since then?
Fisher: Just experience. And I want to make money, be more aggressive, you know? Before, I just, I just I didn't do as well. Like, when I first started, I was kind of like, I didn't catch as much, I didn't do as well, I turned around a lot more. Now, it's a whole different ballgame, you know?
Phaedra: Right. But why's that? I mean, what's the difference?
Fisher: It's experience.
Phaedra: Experience in what? In fishing or . . . ?
Fisher: Fishing, making money, yeah, sort of like that.
Phaedra: So, it's basically, it's a change in attitude then?
Fisher: Yeah, right. Absolutely. The weather does . . . the weather obviously matters. It's just a matter of, you know, with boats like these, you know, they are small boats but they are safe boats. You know? As far as the weather goes, it's just how much you want to take, you know? How much you want to torture yourself.
Phaedra: Really? So it's just how much pain you can withstand. It's not really life threatening, it's just pain threatening.
Fisher: Yeah. [*laughter*] That's how I kind of look at it. I mean, it's not how some other guys think, but there's a lot of guys that think like me, you know? I mean, nice days are nice. Really rough days, you get your ass kicked.
Phaedra: And that's what a real fisherman does, right?
Fisher: You know? But, like, I wouldn't say it's life threatening, you know what I mean? Maybe it's because I have the boat set up nice, and I have everything tied in, you know? It's stable, everything is well maintained. Maybe that's why I don't consider it life threatening. You just get, you just get beat up. When you get in after working a day when it's really rough, you feel like you've been working a week, you know?
Phaedra: Right. But if there's good fish out there that makes it worthwhile.
Fisher: Sure. It's the name of the game.

With the logistics of participant observation too prohibitive to truly entertain as an option, I had to solely rely on fishers' own accounts for insight into how NWS forecasts figure into their process of determining "if the weather is going to be fishable." Soon after arriving at Neborough,

CHAPTER SIX

I started spending time on the piers of "Codtown," in search of fishers doing maintenance work on their boat, and I conducted on the spot interviews with all those who graciously agreed to spare some time and talk about the weather with me. Before long, I figured out that my best bet for running into fishers was Fridays because the Codtown Fish Auction is closed Mondays, and there is little point in venturing out for fish that will be too old to fetch a decent price at the Tuesday auction. While Codtown monopolized my attention on Fridays, early in the spring of 2004 I also started interviewing fishers in another nearby port, "Whaletown." The goal was to establish whether fishers' accounts about their relationship with weather information would withstand community differences as well as differences in fishing type and boat size. All in all, by the time I reached theoretical saturation I had conducted thirty interviews in Codtown and twenty-nine interviews in Whaletown. Most interviews involved one fisher, but a small number included up to two more fishing mates. The average interview length was twenty-five minutes, with one interview lasting less than ten minutes and two interviews going just over an hour.

A Tale of Two Fishing Communities: Codtown and Whaletown

One of America's oldest seaports, Codtown and its history are inextricably tied up with commercial fishing. Codtown was founded in the early seventeenth century by competing British fish companies who sent fishers, salters, and a ship's carpenter to exploit the rich cod resources in the area. It became renowned for its fishing industry and immortalized in books, paintings, films, and legend. Today, with the glory days of the Codtown schooners and salted, dried cod long gone, fishing at Codtown is very much a traditional, family-based affair. The commercial fishing fleet consists of a hundred some odd boats—ranging from thirty-five to seventy feet long—that are primarily locally owned and owner operated. Nonetheless, Codtown remains a full-service regional fishing hub whose commercial fishing infrastructure includes an ice company, several fueling services, seven fish processing plants, numerous wharves for offloading, a seafood display auction, multiple fish buyers, two repair and haul-out establishments, several gear shops, and various public and private berthing facilities.

Still, it is not the infrastructure one notices when first visiting the main harbor on a sunny blue day but the small groups of fishers cleaning their nets and chattering in Sicilian, the colorful boats along a bobbing pier made of faded planks, the stacks of empty wire lobster traps reeking of decomposing sea life, the calls of seagulls overhead, the salty breeze, the ocean beckoning beyond the slender coastline. This picture turns more quaintly festive on other Codtown piers to satisfy the requirements for authenticity and nostalgia of tourists waiting to board a whale-watching boat or feasting on lobsters and clam chowder on the verandas of local eateries. In the wee hours of the morning, thick blankets of fog play tricks with the light and the sound to transport you to a magical Codtown past, while in the winter, fishing life slows down to an eerie gray stillness of frozen water, dead seagulls, and creaking metal. No matter the season or reason, the combined effect of its diminutive size, the spectacular natural beauty of its coastal setting, and the surrounding city, which, replete with old mansions and artist colonies, basks in its historic and cultural heritage, makes the Codtown harbor of today well deserving of its erstwhile name as "the beautiful port."

It is debatable whether Codtown is a fishing-dependent community anymore. Light industry and the service sector are fast gaining in importance, and foreign fish imports have taken the place of domestic catch landings for a number of local processing plants. One thing is certain, however: the commercial fishing community is suffering. The culprit, every fisher will tell you, are restrictive government regulations put in place to manage fishery resources. Historically, fishing in Codtown has revolved around groundfish species, such as cod, haddock, and flounders. In the period between 1975 and 2002, groundfish revenues accounted for between 43 and 78 percent of the total revenues for all landings in the port. But the emergency closure of fishing grounds in 1994 marks the start of increasingly intensive regulation of groundfishing by the National Marine Fisheries Services (also known as NOAA Fisheries). That same year, Amendment 5 to the Multispecies Groundfish Management Plan and Marine Mammal Protection legislation called for a 50 percent reduction of the fishing effort over a five-year period by instituting a maximum annual number of working days at sea in a particular fishery. More amendments followed, expanding closed fishing areas and times and further restricting the maximum days at sea from 139 to 88 by 2009. Between 1975 and 1994, the groundfish catch in Codtown averaged sixty-five million pounds per year; in 2009, only twenty-three million pounds were landed.

It is hard to overestimate the toll fishing regulations have had on the fishing community. Long-standing feuds among different fishing gear groups have been exacerbated as draggers, gillnetters, and long-liners are now all competing for the same hard-bottom fishing areas, and efforts to diversify, or altogether shift, to less-regulated species have created new

CHAPTER SIX

hostilities between groundfishers and other gear groups. Meanwhile, larger boats, too specialized to make the transition, have all but vanished from the Codtown horizon. Gone, too, are fishers under the age of thirty. A recent study reports that the average age for draggers is forty-six, with twenty-six years of experience.[8] Even as they complain that their children have lost their sense of pride for the family heritage, fishers say they dissuade young blood from entering the business.

To be sure, the mythology of the brave and fiercely independent sailor who gets going when the going gets tough still very much captures the collective consciousness of commercial fishers. The difference is that decisions on what gear to use, which species to fish, where, and for how long no longer primarily depend on fishers' knowledge of local marine ecology and their economic calculations. What was largely perceived as a solitary enterprise has now turned into a legal tangle, forcing fishers into uneasy coalitions and growing conflicts with the state.[9] All the while, gentrification, so commonplace in contemporary coastal communities, is yet another source of conflict, this time between the old fishing interests that aspire to a sustainable and secure future and new development interests that envision a colorful "boutique" fishery, if any at all.

Less than a hundred miles south of Codtown, Whaletown has an illustrious fishing heritage all of its own. By the 1800s, it sent out more whaling ships than all other American ports combined. In fact, at the zenith of its whaling prosperity, Whaletown had the highest income per capita in the world. Soon thereafter, however, with whales all but extinct and the discovery of petroleum eclipsing whale oil, the industry rapidly declined, and Whaletown became known as a large-scale textile manufacturer. Today, the city continues to draw a substantial part of its living from the sea. It is the largest groundfishing port in the Neborough area and one of the leading suppliers of sea scallops, ranking as a top-grossing fishing port in the country. In 2002, fishers landed $169 million worth of seafood in Whaletown, more than half of which was generated by scallop landings, a record for the city and one of the highest national values ever recorded for a single port. The working waterfront has over seventy processing plants, numerous gear shops and fueling services, an auction display house, two shipyards, two ice plants. And the commercial fishing fleet counts more than two hundred vessels, evenly distributed between scallopers and draggers. Although there are enough small- and medium-size boats, the piers are clearly dominated by large, eighty to a hundred feet long boats. Like other ports in the region, most vessels are owner operated. Unlike other ports, however, the commercial fishing landscape at Whaletown is marked by the presence of four fleet owners who together control about a fifth of the fleet. In contrast to the misleadingly charming ambiance of fishing life in Codtown, Whaletown's harbor is a structure of concrete and steel at the outskirts of a working-class city. No room here for tourists or nature lovers. The air is buzzing with an unmistakable sense of purpose and productivity, and, for once, the seagulls are easy to miss over the deafening noise of people and machines.

Whaletown's fishing industry has also been hit hard by fishing regulations. The 1994 fishing closures affected scallopers as much as groundfishers. Yet, while regulations for groundfishing are becoming tighter every year as the groundfish population surveys by NOAA Fisheries continue to produce grim results, the scalloping industry got a break in 1999 when, in collaboration with Whaletown scallopers, marine researchers at a local university conducted their own video marine surveys and established that scallops have thrived well beyond NOAA Fisheries' estimates since the 1994 moratorium. Temporary fishery openings as a result of these studies led to more openings and a further relaxation of catch limits for the scalloping industry. Today, there is a palpable air of confidence in the Whaletown piers. Brand new vessels have joined the scalloping fleet, and the maintenance dock is abuzz with work long put off for lack of funds. A careful observer notices, nevertheless, that quite a number of the refurbished scallopers are former dragger vessels, sold by their owners into the scalloping fleet. Thus, the happy tale of the striking revival of Whaletown scalloping is drastically tempered by the uneven fate dealt to the rest of the fishing community. And while the plight of its groundfishers may get momentarily lost amid the hustle and bustle of the recovering industrial port, its decidedly blue-collar character and its persisting problems with unemployment and homelessness puts Whaletown more at the mercy of fishing regulations than the diversified economy of Codtown.

It remains an open question to what extent fishers' use of weather information is specifically influenced by such factors as their fishing gear and boat size, the usual demographic categories, or the properties of their information-sharing networks. For the purposes of this analysis, however, which seeks to capture the underlying *logic* behind weather information use, any such differences among the fifty-nine commercial fishers I interviewed only served to confirm for me and further buttress the coherence of the emergent pattern in reported attitudes and behavior. In what follows, therefore, I will be referring to Whaletown and Codtown fishers or to groundfishers and scallopers interchangeably, or make no distinction at all.

CHAPTER SIX

Calculating Fishable Weather

More than one fisher used the phrase "the weather makes my day" to underscore the fundamental link between fishing and the weather. As in other weather-dependent industries, commercial fishers must anticipate future weather conditions to decide on the level of fishing effort and risk they are willing to take. Such calculated risks most often pertain to issues of profitability. Yet sometimes, and unlike most other weather-dependent industries, they also involve questions of life and death, as commemorated by the cenotaphs in Codtown and Whaletown for the thousands and thousands of fishers who have been lost to the sea.

Obviously, what constitutes meteorological risk for mariners differs markedly from inlanders.[10] But it also differs from recreational boaters as well. Commercial fishers are only interested in the wind forecast. In fact, they say they barely pay attention to the forecast at all come summer. Yet, the less they have to worry about the weather, the more they have to worry about low fish prices. Says one fisher, "There's better money to be made between fall and spring in this racket now than there is in the summertime." Different fishers will, of course, have different tolerances for wind and, therefore, seas. Predictably, the crucial factor is boat size,[11] while other factors include the type of fishing involved,[12] sea experience, and predisposition to risk taking. But the new wrinkle in the equation is the fishing regulations, which have made the decision to go or not to go extremely critical and more complex. Overall, fishers are forced to take more risks wherever and however they can, and deciding whether the weather is going to be fishable takes center stage in these calculations. As a Codtown fisher explains,

> We fish around the weather now. The weather and the market. Before we didn't. But now if you fish in a day that's going to blow around thirty knots, you ain't gonna catch as much fish. You go out there, and I lose seven, eight hours just fooling around before I realize it just ain't going to happen. Seven, eight hours I'll never get back. So the weather right now is really important to us.

Instead of heeding an ominous forecast and staying home, some will consistently choose to go fishing through the bad weather in the face of an impending area closure or exactly because bad weather will drive fish prices up. Even if they are unintentionally caught in bad weather, fishers are now more likely to stay, "get thrown around, puke, and then continue fishing" rather than return to port. The following extreme example comes from a pair of Codtown fishers, who, as they proudly advertise, "are known for going out in rough weather when a lot of people don't go."

We're catching mackerel in the middle of last January, and we're getting eighty cents a pound for the mackerel, we're getting a couple of thousand pounds a day. So I went over [to the owner of the Fish Auction] and I asked him, I said, "Hey, I can get a lot of these, is the price going to be high?" And he said, "Yeah, the price is going to be phenomenal, keep them coming in." The next day, a northeast storm hit, it blew forty-five to fifty knots in the afternoon. When we started out, it wasn't too bad, maybe twenty knots. Blinding blizzard, we stayed and brought in eight thousand pounds of mackerel thinking we're going to get eighty cents [a pound] for them. We had a lot of other fish, too, the boat was almost sunk. . . . It was such a bad storm, all boats from everywhere came in [to the harbor]. It blew sixty knots. Huge factory trawls from everywhere came in, everybody got caught in the storm because it wasn't predicted. They all came in, there was two million pounds of mackerel ended up on the auction. I got one cent. So I went and complained, I said "I almost died out there, we're getting one cent for eight thousand pounds of mackerel? We thought we made some money. We made a decision to go fishing in the storm because you told me the price was going to be high." And they threw me out. I'm not allowed to sell fish there again.

Since then, these fishers have found other buyers and "it works better now," they say. As for that episode, they admit the NWS did predict gales, "but they didn't predict fifty knot winds. We wouldn't have went out if they said fifty knots, absolutely not."

This risk-prone behavior is exacerbated by the fact that years of fishing regulations have resulted in crew layoffs and shift rotations in order to offer deckhands an adequate share. Many owner operators on smaller vessels now fish alone for some or all of the year, while some larger vessels are operating inshore with skeleton crews of just two to four because they cannot afford to work with a larger crew or to fish offshore for any extended periods. Overall, there has been a drop in the number of deckhands from a high of eleven to just two to six. Reduction in crew size has been accompanied in turn by longer trips at sea to reach open fishing grounds, increasing the work load and stress on the remaining crew and making it much more difficult to find good crew for shorthanded vessels in the first place.[13] And it has meant less manpower to deal with emergencies at sea. Yet one also sees many small boats, designed for coastal and not offshore fishing, venture into the same outer waters in search of groundfish.[14] For these day boats, the length of the workday has increased considerably, and it is common for trips to last eighteen to twenty-four hours, primarily because of added steam time to reach open fishing grounds. Fishers, often working alone, are more exhausted than ever and more vulnerable to injuries and accidents, especially in adverse weather conditions.

A tragic illustration of the above is the capsizing of the Whaletown scalloper *Mayday* in the winter of 2004, one of the worst marine accidents in the Neborough area in the last thirty years. Of the six-man crew, only one lived to tell the grim tale. The 75-foot *Mayday* was built in 1979 as a trawler-scalloper but had undergone a number of structural modifications over the years to accommodate a 13-foot-wide scallop dredge off each side. No stability reviews were ever conducted, but none were required. On that fateful day, on their fifth day out at sea, captain and crew were watching the weather because the NWS was predicting gales. They were located in a scalloping area only open between November and January. It is important to bear in mind here that, according to the fishing regulations at the time, fishers who entered these waters and left before completing their allotted catch for any reason, including bad weather, had to forfeit the equivalent of about three thousand pounds of scallops, worth roughly $18,000.[15] As the wind started picking up, the captain decided to tow the dredges a couple more times "to reach our [fishing] limit and then jog home through the weather," as he communicated to a nearby scalloper. Suddenly, a wave threw the boat on one side and water started quickly filling the main deck. In vain did the captain try to dislodge the dredges off the sea ground. Trapped on the main deck, the water toppled the boat over and dragged it under the surface. The Coast Guard investigation that followed names not a single factor but a series of events that led to the tragic accident. But the local House Representative demanded an immediate reform of the fishing restrictions that compelled fishers to remain at sea during adverse weather conditions: "Nobody can say with any certainty the capsizing wouldn't have happened," he said to local media, "but we shouldn't put our fishers in the position of having to consider staying out in inclement weather because they are concerned about losing money."

In 2010, Amendment 16 replaced the days-at-sea system with a sector-based management system where fishing quotas are specified in pounds of fish caught. Although not without its own share of controversy, the new policy aims to alleviate the inefficient and dangerous fishing conditions promoted by the previous program, give some of the control back to fishers, and encourage cooperative harvesting efforts.

Making Sense of Weather Information

Without exception, fishers said they primarily rely on the NWS for information about whether the weather is going to be fishable, many of them even giving me an unprompted demonstration of how they switch to the weather radio band channel while on board. All commercial fishing boats are required by law to carry a VHF marine radio in U.S. waters for communication and safety purposes. Seven VHF frequencies have been assigned to the NWS weather radio, broadcasting on a continuous cycle the latest local land and coastal marine forecast, the adjacent offshore forecast,[16] real-time buoy observations, and tide information. Beyond the NWS, the weather services fishers consult vary considerably and can be summarized according to the following rule of thumb: nearly everyone will watch the weather on their local television stations, but this is an inconsequential act for most, "like if I'm home having supper watching the news." Fishers who have cable television will tune into the Weather Channel over local television, while fishers who have Internet prefer to go online rather than watch the Weather Channel. In addition, a lot of fishers have a VHF radio at home, or they routinely call into the Neborough office hotline for the latest recorded marine forecast. What perhaps is less obvious is the logic behind this preference hierarchy. The perceived accuracy of the weather services in question is largely a moot point and does not determine their order because, in fishers' eyes, it is only the NWS that provides a marine weather forecast proper:

Fisher: I also listen to the weathermen on television, but they don't say much about marine weather.
Phaedra: So why do you listen to them then?
Fisher: I don't listen to them. I only listen to NOAA.[17]

The decisive factor, rather, is the extent of available weather maps. Here is what a fisher with limited access to the Internet had to say:

It's a real vague forecast, the Weather Channel is, it's not really that accurate, but you might be able to kind of plan in your mind what you're going to be doing in a few days, you know? The thing is, their map, with the highs and lows, is accurate. So, you can just look at that. Get a long range kind of idea about what's coming up the coast, stuff like that. The day before, you know, I kind of, I'll look at it all different ways, you know? Then I put on the VHF marine weather. . . . The Weather Channel, I look at it because I can look at the map, see the fronts coming, and interpret it myself. But, I

CHAPTER SIX

mean, I don't really judge when I'm going to be doing fishing by the Weather Channel. You're afraid it's going to be windy but you really want to go, then you watch the Weather Channel. Their wind predictions are a lot lighter. They don't have the buoy reports so, you know. But if you really want to go, you can say, "Oh, it's not going to be that bad." You can tell the guy going with you, too, that it's not going to be that bad.

In effect, what fishers are looking for when they are not listening to the NWS weather radio is more weather information, not a better forecast. It is quantity they are after; quality is a given since "they get all their data from NOAA anyway." Characteristically, the great majority of fishers say the biggest improvement in weather services in the past twenty years has been the amount and frequency of weather information, not forecast accuracy as such. And while all Internet users say they frequent the NWS website for information, many have found a variety of other sites that sport what they consider more user-friendly and insightful weather maps.

Yet fishers do not rely on weather services alone for information. Unlike the dearth of marine weather observations the NWS has to contend with,[18] they have at their disposal a plentiful network of experienced weather observers: the commercial fishing fleet itself. And they take advantage of it as often as they can—the night before, at the pier, en route to a fishing ground. As the following interview excerpt vividly illustrates, the importance of this informal weather network cannot be overstated.

Fisher: We got caught in Hurricane Name and almost died. This was, I forget what year Hurricane Name was, I think it was back in the early nineties or late eighties, and at the time, we went 110 miles out because we just couldn't find any fish in here. So we packed on extra ice, packed on extra fuel, and it's hard to do in this small boat. I mean, on the way out with no fish, we're loaded. We were told Hurricane Name was going on the other side of Florida, was going to hit Louisiana and break up over the land. We went out. Now, we're out there 110 miles, and our little radio can't get the weather. So, the third day we wake up, and there's fifteen foot seas, it's blowing twenty-five miles an hour out of the southwest. It was too rough to haul in all our nets, so we said we'll just stay in the bunk for the day, maybe tomorrow will be better weather. Just by chance, later on that afternoon, a boat was going by. He came over, I knew him. He said, "What are you doing?" I said, "We're waiting for tomorrow." He said, "No, this is the start of the hurricane, you got to go home." And I thought he was kidding, I thought he wanted our fishing spot or something, but after a few minutes talking, he started getting mad and serious and saying, "If I've got to come on your boat, kick your asses, and drive it home, I will. It's a hurricane, you've got to go.

Phaedra: Wow, you guys were lucky!

Fisher: Yeah, we were going to stay. They said before we left it was going on the other side of Florida, it was going to break up in Louisiana. That's when it was off the tip of Florida. It came right up the coast, and there was two or three boats that stayed, and they all died. Bigger than us. They all . . .

Phaedra: Why did they stay? They didn't believe the other guy?

Fisher: No. Well, he didn't talk to them. They were from [another state]. Just a couple of boats.

Phaedra: So if you didn't know the guy, you would not . . .

Fisher: We would have been out there, yeah, we were going to stay. And we had a horrifying ride home . . .

As fish scarcity and fishing regulations drive competition to unprecedented levels, there is some evidence of a breakdown in trust and information sharing among fishers (Thorlindsson 1994, 332; Gazelius 2007; but see Ramirez-Sanchez and Pinkerton 2009). And if fishers are sharing less information about good fishing locations, this means that they are not sharing information about the weather as well because that would give away their fishing spot.[19]

To be sure, personal observations are still considered to be the most reliable source of weather information. Much like weather forecasters, fishers always keep an eye out for the weather: "Looking for birds flying high, and looking for the clouds, how fast they're moving, you know, dark horizons, all kinds of stuff like that." Smaller boats especially that do not venture far off the coast can comfortably suspend the decision of whether to go fishing or not until they can make an assessment on the ground.

If it sounds fishable I go out. If it's rougher than they said, I can make a decision at that time. The ultimate thing for me is I'll go out to the front of the harbor, and if I can't go beyond the breakwater, not a good idea. Basically, I'll look at everything and see but, ultimately, it comes down to your personal judgment. I mean, it could be blowing and you could go, "Well, it's blowing, let's go see." Then you get out to the mouth of the harbor, there's an eight-foot swell going, you're like, "I don't think so." But sometimes you'll hear small craft warning, you'll go out and you go, "Where? This is beautiful. Is it supposed to be at two o'clock or four o'clock?" You're out all day and there's nothing.

As for the predictive value of personal observations, fishers still heed the old weather lore, and, like their fathers and their fathers before them, they pay attention to such natural signs as the color of the sky, the flying of seagulls, or the appearance of a sundog. And they are convinced that, for example, "if you see the seagulls flying way up high, you know you've got wind coming." "But," they are quick to add, "you don't know if it's going to be thirty knots or fifty."

Indeed, commercial fishers' relation to the weather and weather information bears a profound resemblance to that of NWS forecasters. Just like them, fishers voraciously consume information from multiple sources to make a decision. And just as NWS forecasters are primarily concerned with how to best make do with the available forecasting tools, not with the forecasting tools as such, so too are commercial fishers primarily concerned with how to best make do with the meteorological tools at hand, one of them being the NWS marine forecast, and not with the NWS marine forecast as such.

> Phaedra: But if, like you said, they are usually only thirty-five percent accurate, why do you listen to the NWS forecast then?
>
> Fisher: It's something. It's better than just guessing. I mean, it's something. They give us an idea.
>
> Phaedra: It sounds almost like guessing. It's less than fifty percent accurate, so you might as well just flip a coin.
>
> Fisher: No, we'll take their input, we'll take the input from the buoy reports, and we'll take our own knowledge, and pull it all together to make a decision. Maybe like thirty percent from them, thirty percent from the buoy reports, forty percent from us, and we got a system that works.

What is more, commercial fishers' weather calculations, locally circumscribed though they may be, rival in sophistication those of Neborough forecasters: fishers have access to more real-time weather information, their expertise on local marine weather conditions is unparalleled, and their lifetime of experience makes up for the meteorological skill they are lacking.

Ultimately, NWS calculations of meteorological risk and fishers' calculations of meteorological risk stand, if not in an orthogonal, certainly in an oblique relation to each other. Fishers' concerns about fishability are hardly addressed by the conservative "safety first" NWS message. In fact, while I encountered no direct evidence to that effect, it is arguably the case that the more confident a fisher is with his forecasting skill, the more he would be inclined to wish for a "busted" NWS forecast so that he can reap the benefits of the increase in fish value. What may, therefore, at first sight appear as a failure to adjust to the sound advice of the "experts" is often actually not a failure at all but an eminently reasonable decision based on additional, compelling, context-driven considerations.

Such considerations, however, are conveniently hidden from NWS forecasters' point of view. The weather information requirements they imagine fishers to have are not only far more straightforward than reality but also far more compatible with those of other imagined marine forecast users. Yet, without somehow subsuming the complexity of their elusive audiences into a contrived lowest common denominator, how else might they be able to distill atmospheric complexity into a workable definition of sensible weather?

Opinions about the NWS

Perhaps counterintuitively in view of the above, commercial fishers are in many ways the ideal NWS public. What makes them so is precisely their meteorological sophistication. Thanks to their continuous exposure to weather forecast services, they know to stay tuned for updates, and they know how to properly listen to the forecast and gauge forecasting trends:

It seems like they have a bigger wind speed window in the winter. They'll say twenty to twenty-five in the spring and the summer, but they have a ten mile window in the winter. So obviously they have trouble, they're covering themselves when they take a wide range, that means that they're not sure. And we know this. Once you hear them say, you know, "fifteen to twenty-five, gusts to thirty," you know that they don't really know what it's gonna be.

Given their familiarity with local weather patterns, furthermore, fishers know how to transform the generic NWS marine forecast into a meaningfully specific one: "I know if it's blowing out of the southwest, it's going to be choppy. If it's northwest, there's a little lee on the bay, just the way the bay is shaped." And they know how to translate the NWS forecast into a better one:

Generally in the fall, what we normally do is, they say ten to fifteen, ten to twenty [knots]—we take them and add them together. Soon as we get off the coast twenty miles, we add them together. 'Cause that's what you're going to, generally that's what you're going to get. See, in the summertime all the wind's near the shore, and the wind's not off. But what happens [in the fall] is the wind gets stronger, 'cause you have the wind coming from the shore, and it's hitting the water, different change of water temperature and the wind picks up.

CHAPTER SIX

Most importantly, their meteorological savvy and experience have given commercial fishers a realistic outlook about what to expect from the forecast in the first place. They understand that NWS forecasters "are doing the best they can with what they are working with. I mean, it's the weather, it's Mother Nature. You can't predict Mother Nature a hundred percent. When NOAA can start generating the weather, then I'll complain when they can't forecast it."

In this light, it should come as no surprise that the majority of Codtown and Whaletown fishers I interviewed gave high performance marks to NWS forecasts. When asked to express NWS forecasting accuracy in percentage terms, their answers hovered around 80–85 percent, with the highest outlier being 99 percent and the lowest 25 percent. Consistent with their appreciation for the challenges facing the weather forecasting task, fishers allow for a five to ten knot accuracy window: "I mean, it can be a little bit of guesswork but it's enough to prepare yourself for the weather. Because if they're basically five knots more or less does not make any difference. The indication that there is wind coming, it's good enough." And they expect forecasting accuracy to deteriorate somewhat in the winter months because "it's a lot more difficult to predict the weather in the wintertime around here." In fact, they expect forecast accuracy to deteriorate the further out in time the prediction. Says a Whaletown dragger, "It depends on how close you're looking. I don't really count on it for four days out. I'd give it eighty percent for two days in advance, and if I'm looking for four days in advance, then I'd give it about fifty percent. Basically, I don't believe anything after two days. But it gives you a rough idea."

Thus, it is exactly because commercial fishers look so closely at the weather that they know not to look too closely at the accuracy of the weather forecast. By extension, because they are not looking too closely, NWS forecasts are deemed "good enough." Importantly, however, the opposite is also true: because NWS forecasts are good enough, fishers are not looking too closely. Characteristically, when pressed for more concrete answers as to the "what," "where," and "when" of inaccurate forecasts, the great majority could not identify any explicit inaccuracy patterns, and there was no clear consensus among the few Codtown or Whaletown fishers who did. This, I should add, was much to the discontent of Neborough forecasters, who were pumping me for information and feeding me with specific questions to ask. In the same vein, when asked what improvements they would like to see in NWS services, most fishers were interested in better timing and availability, not accuracy, of weather information and forecasts, while an even greater number had nothing to say on the matter.

To be sure, fishers' exasperation with the NWS when caught in unexpected weather comes through very loud and clear:

> When you go out thinking it's going to be good weather and all of a sudden you're getting the shit kicked out of you, excuse my language, then you're saying, what's this guy talking about, you know? It's bad, because he's saying it's going to be, say, northerly wind, right? So I'll go to a spot where I'll be fishing, where I'll be backing stern-to [sea], which side-to is lousy, backing stern-to is good. Now I get to the spot, you told me it's going to be northerly wind, I get to the spot, it's northeast. Now I'm in the wrong spot for northeast, I should be way over there where I can work northeast southwest and it comes perfect. Now you've fouled me all up. By the end of the day, you're drunk, and you're hurt, and you're all beat up from going side-to.

And, as one would expect, disgruntlement over a missed forecast quite often finds comic relief at the expense of forecasters. Says a Whaletown scalloper, "I know that they have their hands tied and they only have so much money; they can only work with what they have. We don't blame them, we just talk about them to get out frustration. [*laughter*] We used to say, I'd like to have that guy strapped to the bottom of the boat now so he can see what twenty to thirty looks like. Up close and personal."[20] Still, in the end, fishers readily concede, "It's not the weatherman's fault. The one who has the ultimate responsibility is the owner or the captain. It's usually they want to work, and they want to go, and they get themselves in situations, and it's too late." And while the widely circulated quip, "if you listen to the forecast, you'll never go fishing," has ostensibly been coined by fishers to poke fun at the typical conservatism of NWS forecasts concerned with "public safety first," it also highlights the precarious conditions fishers must by definition endure if they are to make a living out of the sea. Fishers' frustration over a missed NWS forecast is always tempered by the acknowledgment that managing the weather is fundamentally their job: "The wind comes up, either we turn around and come in, or we don't turn around. It happens. I don't blame anyone. It's our daily business. It's what we do."

None of the Whaletown and Codtown fishers I spoke with—not even those involved in a weather-related marine accident, who otherwise expounded on how the severe weather that caused crew injuries and gear damage had not been forecast by the NWS—has ever called the Neborough office to complain about a missed forecast. In fact, none has ever seen any reason to contact the NWS at all. The hard realities and macho culture of commercial fishing, coupled with a suspicion of government bureaucracy and sharpened by the recent fishing regulations implemented by a sister agency, work to make interacting with the NWS if not an idle pursuit certainly not a priority for them.

Phaedra: Have you ever called them? If not to complain, to let them know that what they're saying does not reflect reality?

Fisher: Nobody wants to listen to me. No, they'll probably make a note of it, but it's not going to change anything.

Phaedra: You don't think it's worth it?

Fisher: No. It ain't worth my time. I got too many other things to worry about. I got too much going on. I get a headache real quick with that stuff, you know what I mean? Talking over the phone with somebody, and then, and then they put you on hold, and then "please leave a message." I just know that type of business that we're dealing with, government, dealing with the feds or the state or anything like that that has a structure, you know what I mean? Press one, press two, and press three, one of them deals. I ain't gonna deal with that.

Phaedra: So has it ever crossed your mind and then you dismissed it, or . . . ?

Fisher: It's crossed my mind. Give them a piece of my mind.

Phaedra: Do you remember any specific occasion when that happened?

Fisher: No, but I just know it wasn't the right weather report, and I got my ass kicked, and I says, "That guy, I'd like to give him a piece of my mind." And you think you're going to do it, you say you're going to do it, but you just never do it. Because nothing's going to change, I think. I could be wrong.

Phaedra: So would you say these forecasting errors are avoidable or not?

Fisher: Well, maybe they're not avoidable, but don't state something if you're not sure, you know what I mean? One time I'd like to, you know, hear him say that "We don't know what the Christ is going to happen today. Youz guys just go, we don't know what's going to happen," you know, because it's almost the same thing. You'll go out there, the guy says it's going to be ten to fifteen, and it's blowing thirty. It's the same thing as them saying, "I don't know what's going on. Play it by ear. Take your own chance." Then it'd be more truthful.

Whose Weather Is It Anyway?
Reprise on the NWS and Its Publics

Based on their core mission, NWS forecasters must tailor their expertise to appear both objective and useful, sophisticated yet conveniently practical. The NWS forecast must first and foremost be actionable. To paraphrase from the quote above, forecasters' challenge is how to indeed state something even if they are not sure. It will not do to avoid issuing a forecast or warning or, even worse, to publicly admit that tomorrow's forecast is likely to be wrong. In fact, any expressions of uncertainty are actively discouraged by the NWS administration, ostensibly in the name of clarity and public safety. Yet also missing from NWS products are quantified expressions of certainty. Explicit probabilistic statements are only to be found in precipitation, hurricane, and climate forecasts, while the presence of a range instead of a discrete value in temperature, wind, and wave forecasts denotes spatial and temporal variation as well as uncertainty. And so, despite the resounding consensus among forecasters that the marine community exhibits a high level of meteorological sophistication, marine forecasts, much like land forecasts, are deterministic yes/no forecasts. For an enterprise whose hallmark is the ability to tolerate uncertainty, the propensity to publicly dismiss uncertainty even while making pronouncements about the future is a rather remarkable, if understandable, practice.[21]

NWS forecasters' challenge, how to state something even if not sure, can be fruitfully seen as a special case of what I have defined as the problem of expertise—namely, how to relate actionable information. In deep uncertainty settings especially, one may reasonably expect that experts' constructs of "the public" will conveniently also provide a solution to the challenge of how to state something even if not sure. Two separate yet interrelated conceptualizations of the public appear to serve this role in the case of NWS forecasters: that people *want* a deterministic forecast because they have to make a decision, and/or that people *need* a deterministic forecast because they do not understand (that they do not understand) probabilistic reasoning. By rendering expressions of probability counterproductive to public service, this narrative—either in its first, weaker version or the more common second one—gives forecasters license not only to refrain from numerically or verbally specifying the uncertainty surrounding a given weather event but also to state something even when not sure. Notably, it formed the basis of forecasters' resistance to the 1965 introduction of a "probability of precipitation" statement in the NWS forecast (Murphy and Winkler 1974, 1979). And, forty years later, it was Neborough forecasters' stock response whenever confronted with the suggestion that conveying the uncertainty associated with a given weather prediction might actually be valuable to their audiences.

One might be inclined to argue perhaps that the practical meteorological knowledge, or *métis* (cf. Scott 1998, 311), of commercial fishers all too easily undermines a model of forecast use that leaves out the reflexivity and identity of forecast users. Yet there is growing evidence to suggest that NWS forecasts are also ill suited for the weather-related plans of far less meteorologically savvy users. This is certainly borne out by the brief survey of 110 recreational boaters I conducted during the Neborough marine spotter training sessions in the spring of 2004.[22] Although the majority of these recreational mariners reported relying nearly a hundred percent on the NWS forecast, they also said they would prefer a probabilistic over the current categorical forecast. Furthermore, despite their markedly more passive relation to weather information, most recreational mariners offered suggestions for NWS improvement that had to do with quantity of information—for example, more frequent updates, more localized forecasts—and not accuracy, thus echoing the attitudes of commercial fishers.

And by no means does the dynamic use of weather information end with mariners. A series of studies show that most forecast users, from building contractors and farmers to homemakers and senior home residents, do not take a weather forecast at face value but instead tend to "keep their own counsel" as to its true significance, adapting the forecast message to their daily plans and routines rather than vice versa (Sanders and Westergard 2002; see also Westergard and Sanders 2000). This is not because they mistrust weather services. In over ten years of research, I have yet to encounter someone who actively mistrusts the NWS in particular or weather services in general, although of course we all enjoy a good laugh at the weatherman's expense. Rather, it is because their concerns cannot possibly be contained in the one-size-fits-all risk management advice dispensed by the NWS.

To anticipate a potential objection to this argument I would like to insist that not heeding a NWS forecast versus not heeding a NWS warning is a difference of degree, not of kind. At their core, *both* responses consist of the same series of nonlinear steps that, at a minimum, include hearing the NWS message, judging that the message is credible, confirming that trusted others are heeding the message, confirming the existence of a meteorological risk, personalizing the prediction to one's situation, determining whether protective action is feasible, determining what action to take, and actually taking action (cf. Sorensen 2000; Mileti 1999, 141; Lindell and Perry 1992). All of the intervening steps between hearing the NWS message and taking appropriate action are necessary for the NWS to successfully protect life and property. Yet only one of them is contingent upon trusting the NWS. Most of the response process hinges on the extent to which the NWS message can be meaningfully contextualized to one's particular circumstances.

As the level of meteorological risk and the concern for public safety increases, so too does the zeal of the NWS to control the risk mitigation process and hence to compress and streamline the steps between someone's hearing of and acting in accordance with the NWS message. Certainly it is worthwhile to pursue ways to engrain better warning response habits and try to circumvent some of the cognitive resistance to behavioral change. Thanks to the small but growing body of work on the social effects of weather information as well as scholarship on natural disasters more generally,[23] there are already concrete measures to be taken along this line. But any such efforts will be off the mark if they continue to be premised on the assumption that the public mistrusts weather services.

Adopting a frame of reference that recognizes the agency of forecast users, however, would be tantamount to a commitment by the NWS to adopt a distinctly different approach to the management of meteorological preparedness and risk mitigation. It would constitute an acknowledgment that NWS forecasts and warnings are, at best, only one among several other pertinent pieces of information that users need to take into account before making a weather-related decision. Yet, policy obligations to not compete with the private sector aside, NWS calculations of meteorological risk can accommodate a rather limited set of local attributes and constraints, even when already restricted to a circumscribed geopolitical domain. Anything more, and constructs of the American public are on the verge of becoming far too complex to remain workable. To be sure, these local attributes and constraints, social as well as geographic, should still be based on systematic empirical measurements and observations to be of value. Beyond this critical prediction baseline, however, the value of NWS forecasts and warnings surely hinges on the extent to which they are able to empower users to arrive at a scientifically informed decision that best suits their own needs. And this translates into forecasters' ability to somehow also communicate, in a relatable and authoritative manner, exactly how confident they are in their predictions.

The academic community certainly seems to endorse this position. There is a growing influx of research that from a variety of perspectives and with a variety of solutions in mind urges for the necessity to communicate forecast uncertainty to the public in order to actually preserve credibility, legitimacy, and trust.[24] Indeed, according to an NWS-sponsored study, the highly detailed, and therefore highly fallible, graphics of the Digital Forecast Database not only perpetuate but actually exacerbate user mistrust (Westergard and Sanders 2000). It is therefore recommended that the NWS move toward forecasts and warnings that enable users to envision alternative meteorological scenarios, thus providing them "with a more scientific basis for second guessing forecasts (as they now do and will continue to do, in any case)" (Sanders and Westergard 2002).

The above study and the forecast prototypes that were developed in the course of it have since been abandoned, the NWS administration purportedly growing resistant to its operational recommendations (personal communication). Instead of working to provide forecast users with more contextualized, open-ended, probability-intensive information as per their true requirements, the NWS appears to continue to invest all of its resources into improving the Digital Forecast Database, which has since moved from a 5- to a 2.5-kilometer spatial resolution. This strategy is entirely in line with the recent effort to brand the NWS as a wholesaler of weather information and in the process keep the People's Weather Service alive. As such, it has much to recommend it. The Digital Forecast Database can potentially be of great value to other "big data" organizations, organizations that neatly map onto those users whom the NWS recognizes as its "partners" and "customers."

Improvements in the accuracy and precision of NWS forecasts, however, are fated to remain largely unnoticed by the American public, just as they have in the past, as long as they are not accompanied by improvements in their social relevance. As for those users who, like commercial fishers, are cognizant of NWS improvements in forecast accuracy because they directly consult the NWS forecast, they are very appreciative of NWS forecasters' efforts but seem hardly as interested in the accuracy of the NWS forecast as they are interested in the amount of information contained therein. In any case, they still also turn to other sources for information about the weather, and they proceed to decide on their own, like everyone else, on the course of action that is best suited to their situation.

And so, the publics of the NWS will continue to look past the NWS forecast for additional information as they make their weather-related plans. And for their part, NWS forecasters will continue to resort to functional—albeit, one hopes, *empirically grounded*—constructs for the weather information requirements of the general public as they deliberate on how to best protect the life and property of the entire nation. Still, to the extent that the NWS continues to assume its weather advice should not be second-guessed by its publics, it will be protecting the life and property of imagined lay persons.

SEVEN

Toward a Sociology of Decision Making

Pray don't talk to me about the weather, Mr. Worthing. Whenever people talk to me about the weather, I always feel quite certain that they mean something else.
OSCAR WILDE, *THE IMPORTANCE OF BEING EARNEST*

Chapter after chapter, the discussion thus far has exclusively concentrated on meteorological decision making. This was intentional—a narrative strategy to allow readers to immerse themselves in the rich complexity and vivid detail of the NWS case and, in so doing, to envision, understand, and ultimately translate into their own terms the world and practice of weather forecasting (cf. Abbott 2004, 30–31). Through a recursive combination of thick description and theoretical engagement, I have endeavored to bring out the multifacetedness of the process of NWS forecasting qua decision-making action. It is thus my sincere hope and expectation that talking about the weather, as it were, has proven insightful to scholars of meteorology and scholars of decision theory alike. Yet, from the start, the ultimate objective of this exercise has been to put forward a preliminary framework for a sociology of decision making. Instead of a conclusion, therefore, this chapter serves to first elaborate the proposed framework and then test its analytic usefulness in two other decision-making fields—medicine and finance.

Let me begin by reiterating my main thesis, abstracted from the specifics of the NWS case: *Decision making is a fundamentally practical activity that relies on available heuristics, techniques, and resources—as determined by both the objective at hand and the evolving material and symbolic context of action—to fashion a provisionally coherent solution to routine and nonroutine challenges.* I have made an attempt to schematically represent this conceptualization of the decision-making process in figure 1. The unit of analysis is the decision-making task at hand. Note the distinction between the Decision-Making Environment and the Context of Decision Making. The former defines the more or less institutionalized realm of decision-making action, whereas the latter defines the particular empirical situation in which decision making actually occurs. Accordingly, during a decision-making episode, one's stock of knowledge is brought to bear on the objective at hand and helps initially frame and specify the context of action. But it is within the microcontext of action and the human and nonhuman others populating it that decision making takes form first and foremost. Which humans and nonhumans are going to be deemed useful resources and which heuristics and techniques applicable is thus going to be determined in and through the ends-means practical act of making a decision and can never quite be settled in advance. As such, decision making requires remaining tactically alert to the evolving scene and ongoing contingencies of action. The practicalities of overcoming a perceived challenge open up a space for improvisation, demarcated through one's interactive engagement in a given objective via the various material and symbolic resources at hand and a set of guiding heuristics and techniques. Taking place, then, at the coconstitutive interface of one's applicable stock of knowledge, available resources, and a progressively concretized objective, decision making materializes as the practice of assembling, appropriating, superimposing, juxtaposing, and blurring of available information into a provisionally coherent solution—namely, as the art of collage.

Through the case of NWS forecasting, I have already sought to flesh out and elaborate this basic decision-making model along a number of dimensions: the extraordinariness of the decision-making challenge, the time horizon of the decision-making task, the precariousness of the decision-making situation. I turn to each of these dimensions more systematically below. Several other critical dimensions, such as the expertise level of the decision maker or the regulation of the decision-making environment, must remain beyond the scope of this analysis for lack of adequate empirical evidence. I will, however, be elaborating the proposed conceptual framework along two additional dimensions when I proceed to examine its applicability in the fields of medicine and finance: namely, the interventionist scope of decision making and the performativity of decisions.

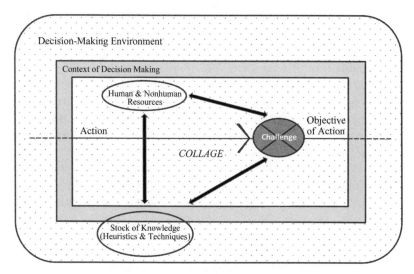

Figure 1. Conceptual schematic of the decision-making process

Decision Making and Practice

In principle, the entirety of our daily actions and interactions—from the moment we wake up in the morning to the moment we drift out of consciousness at night—is the result of innumerable decisions. Yet, for better or for worse, the vast majority of these decisions have become unrecognizable as such for all intents and purposes, reduced to spontaneous, unreflective reactions to the empirical context of action (Schatzki 1996). Just as we are able to seamlessly move through an array of tasks still half asleep before our first coffee, so too are we able to mindlessly but competently carry out the multitude of responsibilities that make up daily life. It is on this most remarkable, powerful empirical feat—namely, the sedimentation of decisions into nondecisions—that social theories of action typically draw to analytically tackle the duality of structure and agency (cf. Bourdieu 1977; Giddens 1984; Sewell 1992). Notwithstanding a theoretical commitment to a dynamic conceptualization of practical activity, it is thus routinized instances of practice that typically get foregrounded in the extant literature.

In this book, by framing my analysis of NWS forecasting operations as a study of the *practice* of meteorological decision making, I have endeavored to illuminate the entire spectrum of decision making in action. Toward one end of the spectrum, resides routinized, rote-based decision-making behavior, the kind typically encountered in most accounts of practice.

CHAPTER SEVEN

The best illustrative example of this decision-making behavior in the NWS case would be producing days five through seven of the forecast. As we have already seen, while laborious and time-consuming, this is nonetheless a highly routinized undertaking. It entails using a single, prespecified resource—that is, GFS-based Model Output Statistics (GFSX MOS) guidance—to populate the forecast grids, and therefore it requires simply adhering to established, clear-cut, and basic procedures. Referring back to figure 1, the empirical context recedes into the background and plays a minimal, if any, active role in shaping decision making. The scene of action has become all too familiar and taken for granted, its features "seen but unnoticed" (Garfinkel 1967, 36). Indeed, other than a situation where parts or the entire GFSX MOS were to go missing somehow, it is hard to imagine any scenario in which producing days five through seven of the forecast could present a cognitive challenge, not to mention necessitate actual meteorological decision making, from NWS forecasters. Forecasting practice toward that end of the decision-making spectrum is thoroughly colonized by organizationally prescribed rote behavior as articulated and embodied in heuristics and techniques for diagnosing and predicting the weather "this far out." Hence, Neborough forecasters remark about the mindless drudgework, not the cognitive strain, involved in carrying out this supposedly particularly uncertainty-laden task.

Toward the other end of the spectrum, decision making acquires the agentic character we readily associate with it. The empirical context looms large here, and it exerts a dramatic influence on action. Invariably, the decision-making challenge is truly extraordinary, such that one's stock of knowledge is deemed unmatched to the task. Schütz often likened his "stock of knowledge" concept to "a tested system of recipes" (Schütz 1944, 502), "recipes indicating how to bring forth in typical situations typical results by typical means" (Schütz 1970, 239–40). This "set of recipes of more or less institutionalized habits" helps us come to terms and master recurring situations, and thus find our "bearings without difficulty in the common surroundings" (Schütz 1970, 82). Occasionally, however, our recipes for action fail us. For a variety of reasons, we come upon a new or unexpected circumstance that proves unamenable to our usual problem-solving heuristics and techniques. In such a circumstance, the particular human and nonhuman materials on hand—for example, human information sources, human skill sources, technologies and tools, natural resources—are vested with an exceptional power to inspire and direct decision-making action. This is not to say of course that just because we have momentarily lost our bearings we are "flying blind," as it were. After all, institutional policies of "what is doable" are sufficiently mutable to provide at least some guidance for resourcefulness and "innovation though recombination" (Clemens and Cook 1999, 448), and recipes for action are sufficiently transposable to become "actualized in a potentially broad and unpredetermined range of situations" (Sewell 1992, 8). Still, when the typical decision-making heuristics and techniques do not seem to apply, it is the empirical context of action that will determine, first and foremost, how the challenge can be resolved.[1] To push Schütz's "cookbook knowledge" metaphor a bit further, when the typical decision-making recipes do not seem to apply, it is the ingredients on hand that will determine what one may make out of the situation.

If at one end of the spectrum resides routinized decision making and at the other end agentic decision making, then the rest of the spectrum is occupied by—for lack of a better word—*habitual* decision making. As already noted, Dewey's distinction between intelligent and routine habits dissolves the structure/agency binary by reassigning routinized and agentic action to the two extremes of habitual action. It is because we tend to see habit as the opposite of agency that we have come, he says, "to eject the habit from the thought of ourselves and conceive it an evil power which has somehow overcome us" (Dewey 1922, 24). On the contrary, he insists, we *are* our habits: "Character is the interpenetration of habits" (Dewey 1922, 38). Hence, habituated decision-making practice is usually far from being routinized, mindless, or unreflective. Neborough forecasters, as we have seen, do not produce the first four days of the NWS forecast unthinkingly, nor do they issue hazardous weather warnings and advisories mechanically. Yet neither does their decision making resemble the agentic behavior at the other end of the spectrum. Both extremes are relatively rare in practice. Rather, I have argued, weather forecasting involves habitual decision making during what are considered routine *as well as* during what are considered nonroutine operations—that is, during both "fair" and hazardous weather conditions. Decision making can be discerningly creative and skillfully efficient precisely because it is fundamentally based on practiced, habitual ways of reasoning and doing.

Even in highly institutionalized environments, such as the NWS, then, where decision-making habits typically come bundled in relatively coherent "habit sets," the process of decision making is hardly ever fully routinized or seamless (cf. Gross 2009, 371). No matter how much administrators may strive for a logically consistent, unified culture of practice, the cultural repertoires decision makers actually draw on are complex, uncoordinated and, at times, contradictory (Swidler 1986, 2001; see also Lamont and Thévenot 2000). Add to this that decision makers are not only equipped with symbolic but also with a host of *material* resources, and cultural repertoires of decision making acquire tremendous range, granularity, and flexibility. The need for "drastic and costly cultural retooling" (Swidler 1986, 277) is hence relegated farther to the extreme of the spectrum. And it becomes possible to see how decision makers may be able to strategically accomplish modest, partial, or incremental changes to their cultural repertoire within a continuum of relatively stable, habitual action.

Decision Making and Time

It is a truism to say that time matters in how we make decisions. For one, time matters in how practiced we become at making certain kinds of decisions. Yet the decision-making process is not only structured by time experienced as recurrence. It is also structured by time experienced as duration. After all, it is constraints in time that, along with constraints in information and cognitive capacity, are mostly to blame for the boundedness of human rationality.

The vast majority of our decisions are prompted by objective and subjective deadlines for action. Different time constraints, in turn, result in different patterns of decision making. In fact, some dual-process theories claim that having, or taking, the time to make a decision activates an altogether different neural pathway in the brain (but see Van Overwalle and Vanderkerckhove 2013). And while cognitive psychologists generally avoid drawing the conclusion that slow or explicit reasoning is inherently better than the default mode of fast or implicit reasoning, their findings on the whole lend credence to the long-standing admonition that one should resist "snap judgments" and "jumping to conclusions."

It might seem, then, that time pressures only further amplify our natural propensity to "unthinkingly" reach for a decision. But this by no means has to be the case, as my analysis of NWS forecasting demonstrates. Less time does not necessarily lead to less deliberation. Neither does more time guarantee a more deliberate and measured decision-making approach. Long before it became possible to systematically probe into its mechanics, social actors have variously endeavored to offset the limitations of human cognition. A first step, whenever possible, has been to stretch out the time frame for making a decision. That may be accomplished by pushing back the deadline, of course; but an alternative, or additional, strategy has been to redraw the boundaries of what constitutes relevant, contextual information outward so as to include more and earlier options for action. Thus, NWS forecasters do not merely postpone arriving at a prediction when dealing with relatively long-fused weather events, such as snowstorms or hurricanes, but they also *precipitate* deliberation by regularly incorporating information from the Climate Prediction Center into their diagnostic process, several days before a weather system would have an opportunity to organize into a threatening storm. "Buying time" is usually thought of as a reactive decision-making strategy, yet it can also be a proactive one—by recontextualizing available resources, objectives may come sooner, and clearer, into view.

Once the time frame for decision making has been established, so too has one's decision-making approach. To be sure, per pragmatist theory, both approach and time frame are likely to be adjusted in big and small ways in reaction to the evolving context of action. The larger point to be made in this connection, however, is that time as such does not drive decision making, because it simply does not exist outside action and interaction. Rather, as Andrew Abbott (2001b 232ff.) has argued by adopting Whitehead's processual ontology to the study of temporality, our concept of time is "prehensive": it takes shape precognitively, triggered by the mobilization of ingrained logics of action and articulated in direct accordance to the concrete problem situation at hand. But if our concept of time is relationally defined and does not preexist the realm of practical action in which it acquires its meaning for us, then the pace with which we reach decisions is highly susceptible, at a deeply visceral level, to the structuring forces of the decision-making environment, not just our biological wiring. Hence, within a given epistemic culture, particular time frames tend to summon up—and in return be sustained by—particular decision-making regimes. And, explicit or not, criteria for what constitutes proper decision making inevitably include expectations about the rhythm, not just the substance, of the deliberative process. Whatever the natural dispositions and limits of human cognition may be, it is the decision-making environment that plays the conclusive role in conditioning how measured and mindful we actually will be in practice.

Although many scholars recognize that the environment—however theorized—plays a critical role in shaping decision making, they tend to fall back to dual-process explanations when it comes to the temporal structuring of decision-making practice. The work of Gigerenzer, Klein, and their collaborators on the mechanisms of "ecological rationality" (Gigerenzer, Todd, and the ABC Research Group 1999, 18) has yet to reach a broader audience, least of all among sociologists. While highly insightful and important, furthermore, even that line of work offers a rather limited view of how the environment can structure decision making temporally because, overwhelmingly, the empirical focus tends to be

CHAPTER SEVEN

on rapid, time-pressured problem-solving tasks. As we saw in the case of weather forecasting, however, recognition-primed decision making and other such "fast and frugal" heuristics are but one aspect of the NWS effort to mold time into an organizational resource and forecasters into poised decision makers. Deliberation within a relatively long time frame carries with it similar scripts for properly enacting timeliness. The aim always is to, as best as possible, bring the empirical context of decision making in line with internal competencies and objectives. Thus, decision makers have been afforded an assortment of heuristics, techniques, instruments, and protocols to help them identify and navigate a recommended course of action effortlessly, efficiently, and timely. Fast-paced contexts will, no doubt, call forth a set of skills and resources best suited for keeping up with the action. But slow-paced scenarios come bundled with an equivalent set of skills and resources, meant to elicit good long-endurance performance. Just as importantly, decision making is kept at a manageable pace not only because actors are equipped to adapt themselves to external time pressures, but also because they are equipped to adapt external time pressures to internal temporal orders that work to punctuate, sequence, and actually slow down the rhythm of action to viable levels.

In short, the environment plays a critical role in shaping decision making precisely because it also shapes its temporal unfolding in practice. This, ideally, transpires through two collateral adaptation mechanisms: over time equipping decision makers with institutionally sanctioned heuristics and techniques, and equipping the empirical context of decision-making action with institutionally sanctioned human and nonhuman resources (cf. fig. 1). Thus, the NWS seeks to cultivate sets of temporally judicious decision-making habits in its forecasters *both* by promoting expeditious meteorological skills and rules of thumb *and* by scaffolding the temporal architecture of a given task onto more or less fixed deliberation structures and technologically hardwired timing sequences.

To be sure, the streamlined coherence of such decision-making protocols rarely survives unscathed in the heat of the moment, no matter how contextually nuanced or flexible they strive to be. Knowing when to strike while the proverbial iron is still hot remains an *empirical* problem whose resolution in practice is made that much more complex by the fact that decision makers often have multiple irons in the fire vying for their attention and/or must work together with multiple other actors, sometimes from distinct decision-making environments. In extreme cases, the competing logics of action summoned up by the microcontigencies of the situation may collide into fatal "temporal failures" (cf. De Keyser and Nyssen 2001, 172–74). In all cases, however, these are problems in need of sociotechnical solutions. My earlier statement bears repeating: whether hot or cold, fast or slow, implicit or explicit, by the time judgments have crystallized into a decision, they have become tempered through particular institutional expectations and mechanisms of sobriety. While some decision-making environments are more effective than others in regulating and supporting actors in their task, all are decisively involved in shaping human cognition, such as it is, into appropriately externalized enactments of reasoning.

Decision Making and Risk

If Anthony Giddens (1991) is correct that the identifying characteristic of modern existence is the preoccupation with controlling time, then one might well construe current decision-making practice as a project of "the colonization of the future" (Giddens 1991, 111ff.). Fate and destiny have, accordingly, been formally replaced by risk and chance in accountings of action, which now is held to be calculable, at least in principle. Given the profound newfound uncertainty and complexity surrounding most decisions, the calculability of risk and of action has been largely entrusted to so-called expert decision makers, charged with assessing and monitoring risk in an increasing number of life sectors. In fact, many such expert settings of decision-making activity develop into "institutionalized risk environments" (Giddens 1991, 118ff.), constituted through and actively courting risky decisions.

Yet, although the decisions of these experts are typically highly consequential, affecting virtually everyone according to Giddens, much of expert decision-making activity is nonetheless habitual or even routinized. To put it in Goffman's terms, although they occur during "on"—and not "dead"—time, most expert decisions nonetheless do not occur during fateful moments (Giddens 1991, 113). The given situation may be highly consequential, but it is not seen as "problematic." Per my reading of Goffman's definition of where the action is, therefore, the situation is not cognitively straining or effortful. And so it comes to pass that, most often, highly consequential judgments present little difficulty for decision makers. In what would otherwise be considered a high-risk situation, they are able to habitually rely on well-tested heuristics and techniques to efficiently carry out the task at hand.

Occasionally, however, situations are not only highly consequential; they come across as problematic as well. During those moments, decision makers can no longer proceed with "business as usual" and become confronted with the fatefulness of their decisions. In terms of the spectrum of decision-making action outlined earlier, decision making is still mostly habitual, but it now relies on heuristics and techniques developed for specific nonroutine scenarios. NWS forecasters who have had to often contend with tornados or with snowstorms have thus acquired an extensive stock of knowledge for dealing with such events.[2] Fateful moments may come about unintentionally, as is the case for hazardous weather events, but they may also be intentionally triggered (Giddens 1991, 114). Either way, the logic of practice in expert settings qua institutionalized risk environments makes them particularly prone to fateful decisions.

It is during those critical, fateful moments, when our usual recipes for action seem inadequate *and* we are confronted with a high-risk situation, that the importance of situational awareness, what I have termed *aesthesis*, becomes most apparent. For, in effect, by biasing decision makers toward an institutionally sanctioned course of action, aesthesis serves to shield and support them against the dread of a fateful error in judgment. Given the propensity of expert settings toward fatefulness, it is safe to assume that inculcating members with an appropriate aesthesis constitutes an essential organizational goal. This inculcation becomes most manifestly and systematically undertaken at an epistemic level, but it is frequently also sustained on aesthetic, pragmatic, emotional, and moral grounds. The objective, as per the NWS case, is to regulate the cognitive labor as well as the "emotional labor" (Hochschild 1983) that decision-making practice requires. In the process of acquiring a keen eye, practitioners inevitably also acquire a keen taste for certain kinds of risk and risk taking that, by effacing the gravitas of competing choices, works to temper and calm bouts of over/underconfidence.

Decision Making in Other Fields of Action

I now turn to decision-making practice in two other expert settings: medicine and finance. Discussion of these decision-making environments will of necessity have to be brief and to the point. Brief, because I could not possibly begin to cover the gamut of decision-making practices in either professional field in just a few pages. There is enough evidence in the extant literature to conclude that the logic and sociotechnical culture of the various specialties in each field are sufficiently distinct to forbid sweeping generalizations. To the point, because to the extent that one can nonetheless still speak of a more or less shared epistemic culture for each field, I will attempt to strategically highlight those aspects of decision-making practice in "medicine" and "finance" that either illustrate the empirical applicability of my proposed framework outside the field of meteorology or analytically extend it along a new dimension.

Medicine: Decision Making and Intervention

At first glance, medicine and meteorology appear more different than alike as decision-making communities: medicine is diagnostic, whereas meteorology is prognostic; medicine is curative, whereas meteorology is consultative; medical risks are personal risks, whereas meteorological risks are collective risks; and so on. But it is precisely such differences that invite a comparison between the two decision-making fields. They offer a unique opportunity to further elaborate the proposed framework for studying decision making in action provided that it can successfully accommodate them.

Let me begin with the similarities. Like weather forecasters, doctors cannot "act at a distance" (Latour 1987, 229), removed from the messiness of the world "out there." As clinician and public health scholar Eric Cassell (1991) has argued,

> the individual doctor is the bridge, the tacit partner of abstract knowledge that makes it work for each individual patient. This function of doctors included (or, more accurately, hidden) in the idea of judgment, is necessary for the care of the sick not because medical science has more to learn—its knowledge is imperfect—but because the *ideal* of scientific knowledge will not work for *this* sick person without the aid of *this* doctor. . . . When we are sick we do not need impersonal knowledge; we require *personalized* knowledge. (133; emphasis in original)

Hence, doctors are forced to carve a liminal place between the laboratory and the field in order to produce personalized, *practically* relevant knowledge. Notwithstanding institutional pressures to scale down the complexity of the patient body into a neat and authoritative "case," notwithstanding "training for certainty" (Atkinson 1984), they routinely must still adapt their decision-making process to the messiness of medical practice. To be sure, the host of diagnostic and therapeutic heuristics and techniques at doctors' disposal are critical for creating a provisional order out of the complexities of the medical situation at hand (Smith and Hemler 2014). Especially when routinization fails, institutionally ingrained habits of reasoning and doing help protect medical practice from chaos (Chambliss 1996, 49–55). Ultimately, however, the inherent ad hocism of medical work resists efforts at rationalization and standardization (Berg 1997; Timmermans and Berg 2003). Medical judgment and decision making are fundamentally pragmatic because they must perforce remain flexible and alert to the evolving empirical context of action.

In fact, based on several ethnographies of hospital life, it would appear that medical decision-making practice may be most appropriately conceptualized as the art of collage—as a process of assembling, appropriating, superimposing, juxtaposing, and blurring disparate pieces of information into a provisionally coherent therapeutic plan. For example, Annemarie Mol (2002, 71) has shown that the multiple, locally constructed ontologies of atherosclerosis are made to provisionally "hang together" into a "patchwork singularity" through a variety of material and discursive coordination practices. Similarly, Alice Street (2011) discusses the medical chart as a fluid, reconfigurable decision-making device that, rather than locking information into a single diagnosis, serves to pragmatically maximize pathways to action. Beyond the question of how information bricolage congeals into a decision (Daipha 2015a), the heuristic of collage thus helps theorize challenges unique to decision-making settings, such as hospitals, where problem solving is typically accomplished in cross-functional teams, representing multiple distinct communities of practice. In such settings, meaningfully integrating information can be especially daunting, even when team members have developed a working relationship, because of differences in terminology, conceptualization heuristics, evaluation techniques, and/or jurisdictional disputes.[3] Yet a series of studies suggest that cross-functional teams need not invest in practices of deep knowledge sharing to be successful. Especially when procedural protocols are in place, team members are able to dispense with transforming local understandings into a shared, common knowledge and instead focus on juxtaposing "their diverse efforts into a provisional and emerging *collage* of loosely coupled contributions" (Kellogg, Orlikowski, and Yates 2006, 38; emphasis added). Thus, in their study of trauma care teams, Samer Faraj and Yan Xiao (2006) show that treatment decision making and coordination are not facilitated through dialogic exchange but through the use of sufficiently flexible protocols that delineate how medical expertise is to be practically distributed among anesthesiology, nursing, and surgery.

Like weather forecasters, doctors are sometimes confronted with fateful decisions. In fact, given the iconic life-and-death questions involved, it is fair to assume that fatefulness is more readily associated with medical decision making than with meteorological decision making in the popular imagination. But, of course, the consequentiality of medical practice is no less habitual or routinized than is weather forecasting practice. Medical education is notoriously long and rigorous, with a strong ethos of apprenticeship that, in the United States, stretches beyond the graduate medical degree into a year of internship and several years of residency training and often also includes some years of fellowship training before board certification can be sought. In hospital settings, the inculcation of medical aesthesis continues on with daily doctor rounds, conferences, and grand rounds. And while a rising number of physicians has been decrying the contemporary decline in bedside skills and the overreliance on diagnostic technologies (e.g., Verghese et al. 2011), there can be no doubt that the institutional regulation of their emotional labor to project a "studied calm" (Groopman 2007, 74–75), especially when in the presence of a patient, is ongoing and profound. Research on medicine's "hidden curriculum" thus consistently finds an escalation of cynicism and an atrophy of idealism and empathy as medical students become socialized into the profession (Hojat et al. 2009).

Like weather forecasters, doctors explicitly rely on temporal frames of reference to structure their decision-making process. Medical practice is divided among (1) "acute" care, encompassing rapid onset and/or short-duration conditions, both mild and severe, (2) "chronic" care, encompassing progressive and/or long-duration conditions, and (3) "subacute" care, a more recent medical decision-making regime, which encompasses a broad range of medical and rehabilitative services that fall between acute and chronic care. But here is where the similarities between medical and weather forecasting practice end and the differences begin. Despite some variation across specialties, the temporalization of medical decision making operates on such a fundamentally different timescale than meteorological decision making—spanning up to 78 hours for acute care, up to 12 weeks for subacute care, and up to a lifetime for chronic care—to make any attempt at a conversion between the two prohibitive. Even more consequentially, the temporality of medicine is so expansive that it has become organizationally unsustainable. Acute, subacute, and chronic care for critical medical specialties is typically carried out in separate hospital or outpatient facilities, staffed with mostly separate medical personnel. The temporal structuring of the medical task does not only result in distinct decision-making regimes but also in distinct organizational units to remain practicable.

But the most crucial difference between medicine and weather forecasting for the purposes of this analysis is their orientation to intervention. Weather forecasting is essentially a consultative field. Despite the long history of efforts to "fix the sky," weather modification remains of dubious scientific value (Fleming 2012; NRC 2003). More fundamentally, the weather is not personal. Neither, therefore, can be any information about it. All NWS forecasters can do to "protect life and property" is provide the best meteorological decision support possible. And the same holds true for the more specific and customized weather services offered by private meteorology. In contrast, the medical field promises to actually help solve health problems or, at least, to manage health risks. Hence the growing distinction in clinical research between "prognosis" and "prediction," where the former estimates the risk of relapse and death independent of therapy while the latter takes into account specific treatments. What, then, does the day-to-day practice of decision making look like in an environment characterized by a teleological, interventionist logic of action? More to the point, can medical practice be accommodated by a model of decision making that foregrounds action in the subjunctive mood?

Importantly, the now conventional model of medical practice as curative medicine is a relatively recent phenomenon. It was not until the 1930s and the discovery of effective treatments for a variety of common diseases that the shift from a "profession of discovery" (Pickstone 1990) to a "profession of intervention" (Cassell 1985, 8) took place. By the time this shift to treatment and therapy was complete, another important shift had also been accomplished: medical decision making had become reoriented toward diagnosis and away from prognosis.

The de-emphasis of prognosis in modern medical thought and training has been well documented, most notably through the work of sociologist and physician Nicholas Christakis (1997, 1999; see also Mackillop 2008). Although not uniform across all medical specialties, this "ellipsis of prognosis" is evident throughout the profession. The pursuit of treatment and therapy—through a diagnosis-based conceptualization of disease—compels doctors to "blithely neglect" making predictions about likely outcomes as a way to cope with the uncertainties of illness (Christakis 1999, xix). Prognosis is now seen as intrinsic to diagnosis and therapy, explicit attention to it has diminished, and prognostic judgments would appear to have been mostly relegated to the predictive probabilities used in evidence-based medicine. In fact, according to Christakis, doctors not only neglect but actively *avoid* explicitly considering predictions when making treatment decisions, let alone when discussing treatment options with patients and their families. Across the profession, prognosis is regarded with anxiety and disdain given the uncertainty and error proneness associated with it, and doctors employ a number of techniques to delay, reduce, or entirely refrain from prognostication in their decision making (Christakis 1999, 75).

And yet, prognostications are ever-present in medical practice. Not only during doctor-patient interactions—although, of course, it is precisely such interactions that give medical prognosis its moral imperative—but permeating all facets of medical decision making. Explicitly or implicitly, prognosis shapes treatment decisions in multiple, often radical, ways. Prognosis is responsible for switching patients from curative to palliative care, of course, but it may also be the deciding factor for the allocation of scarce resources, including doctors' time and effort (Christakis 1999, 34). Research shows, for example, that insurance considerations and clinical judgments about a patient's "salvageability" play a considerable role, above and beyond any diagnostic criteria, in how doctors decide on relevant therapeutic actions (Crane 1975, 40–61; Caronna 2011). Meanwhile, especially in the case of stigmatized diseases, prognosis may also have an indirect effect on treatment decisions, as it can be quite consequential for arriving at a diagnosis in the first place. Doctors may be reluctant to diagnose fetal alcohol syndrome, for instance, in light of the potential impact such a negative label can have on child and family (Armstrong 2003, 122–28). To further complicate matters, because of the expansive temporality of medical care, doctors are called upon to make prognostications both about short-term "results" (typically associated with post-operative and/or hospitalization endpoints) and about long-term "outcomes." While the double presence of the future is hardly exclusive to the decision-making environment of medicine, what is distinctive about medical prognosis is that doctors must typically address this double future *concurrently* because of their mandate to heal. Even in emergency care, it is not out of the ordinary for long-term considerations to trump short-term recommendations, indeed to trump the need for a definitive diagnosis altogether (Daipha 2015b). The collective coordination of a doubly uncertain future serves to rally and integrate practice within and across hospital units toward a common plan of action.

In short, despite its interventionist logic, despite its all but explicit aversion to prognosis, medical practice routinely casts predictions about an uncertain future to settle on practical solutions for the present. Consistent with Dewey's (1922, 322) claims about the centrality of foresight in deliberation, doctors use prognostications as hypotheses, or "procedural means," with which to guide and refine their diagnostic and therapeutic reasoning. Indeed, it is only within a pragmatist framework of action that the overarching role of prognosis in medical practice is made explicit and empirically accessible. Far from undermining it, medical decision making further showcases and elaborates the analytic potential of a decision-making model that conceptualizes situated action as an ends-means continuum.

CHAPTER SEVEN

Finance: Decision Making and Performativity

At first glance, the similarities between finance and meteorology appear to outweigh the differences. After all, both finance and meteorology are prognostic, both are consultative, both financial risks and meteorological risks are collective risks, and so on. But there are also important differences between the two decision-making fields that make their comparison another promising opportunity for elaborating the proposed framework.

Once again, let me begin with the similarities. Much like the atmosphere, the economy is a dynamic system that is highly sensitive to initial conditions. Like weather forecasters, then, economic decision makers must routinely contend with the fundamental uncertainty of "unknown and unknowable" future events (Dequech 2006). They may be intentionally rational actors, but their rational calculations cannot predict the future with perfect accuracy. In fact, it is precisely the inherent uncertainty of economic action that increasingly serves as the basis for the sociological critique of orthodox economics. Rather than attending to the "irrational" principles guiding economic decision making, the so-called new economic sociology has shifted its focus from decision makers' inner motives to the social situatedness of their behavior. According to this new line of critique, "even if it is accepted that actors in the realm of the economy do attempt to enhance their selfish goals," such actors nonetheless "cannot anticipate the outcome of a decision and cannot assign probabilities to the outcome" because they are operating under fundamental uncertainty (Beckert 1996, 804).

By definition, managing risk under fundamental uncertainty is the *business* of finance. Giddens regularly uses the finance sector as an example when discussing institutionalized risk environments and the fatefulness of decisions. Decision making in finance is not only steeped in uncertainty because it is intent on colonizing the future, but it is also highly consequential because it carries collective, indeed global, repercussions. Like weather forecasters, however, economic decision makers are mostly spared the cognitive strain attached to making problematic, fateful decisions. In practice, they can typically rely on well-tested heuristics and techniques to efficiently carry out the task at hand. Even as they highlight the "productive life of risk" (Zaloom 2004), recent studies of finance reveal an institutionalized risk environment of "structured anarchy" (Abolafia 1996, 38ff.) that cultivates in its members pragmatically "liquid" decision-making strategies (Ho 2009) and certainty-constructing emotion rules to face an unknowable future (Pixley 2009). Meanwhile, calling attention to the situational rationality governing economic action, economic sociologist Jens Beckert (2013) goes as far as to argue that what motivates decision making in the face of fundamental uncertainty is "fictional expectations": imaginaries about the future that serve as the means through which purposeful actors can successfully pretend to work toward a concrete objective and thus avoid paralysis or randomness.

Like weather forecasters, economic decision makers are part of an immense knowledge infrastructure—a cognitively distributed, technologically mediated "economy of calculation" (Hardie and MacKenzie 2007). While modular, supple, and ever-adaptive (cf. Edwards et al. 2013, 5), this heterogeneous assemblage of people and things is eminently material. To be workable, information must become technical and corporeal. Nowhere is this fact more evident than in decision-making environments—like weather forecasting and finance—that are tasked with coordinating and processing enormous amounts of data across the globe. Not surprisingly, then, it is in the context of the study of financial markets that the call for a "material sociology" has been most clearly articulated. "The properties of artefacts, technological systems, conceptual tools, and so on are not 'details' that sociological analyses should set aside," writes science and technology studies (STS) scholar Donald MacKenzie (2009, 3–4; see also Preda 1999; Beunza, Hardie, and Mackenzie 2006): "Economic agents . . . are not just 'naked' human beings, nor simply human beings embedded in social networks. Their 'equipment' matters. A trader equipped with Black's sheets was a different economic agent from one trading on the basis of intuition and experience alone. The Black-Scholes model was indeed technical, but it was not a 'mere technicality': it was a consequential part of how economic agents were constructed." Underscoring the paramount, constitutive role of the material in how actors relate to each other and to the decision-making task at hand, a series of ethnographies of finance in the STS tradition delve into the ecologies of trading rooms (Beunza and Stark 2004) and the architectures of financial markets (Knorr Cetina and Bruegger 2002; Knorr Cetina 2003) as a way to capture the process of accessing risk and recognizing opportunity. In all of these studies, just as in my analysis of weather forecasting practice, screens and screenwork emerge as crucial for provisionally cobbling together "the market" out of disparate, overlapping, redundant, and uneven bits of information about it.

But here is where the similarities between finance and weather forecasting end and the differences begin. For one, economic forecasters do not enjoy weather forecasters' rare high-level calibration, as they cannot avail themselves of similarly well-documented, near real-time feedback on their performance. To further complicated matters, economic decision making operates at both considerably shorter and considerably longer time horizons than meteorological decision making, making the indeterminacy surrounding the financial forecasting task and the tendency toward over/underconfidence that much more acute. But the most critical difference between finance and weather forecasting, of course, is that the former is a social science and the latter a natural science. Setting aside the larger, much thornier question of whether it is scientifically possible to predict social events and behavior in the first place, the fact remains that economists, notwithstanding the mathematical sophistication of their models, have thus far not been able to reliably anticipate financial crises, including near-catastrophic systemic risks.

Unlike the behavior of the atmosphere, financial behavior—like all human behavior—is not only agentic but also reflexive: in an effort to successfully speculate market moves, economic actors fill their perceived gaps in information by using their perceptions of other actors as informational inputs. While strategies of "reflexive modeling" are typically a source of correction, moving prices toward their efficient value, they also give rise to cognitive interdependence among supposedly independent actors and may lead to a dramatic amplification of error (Beunza and Stark 2012). This amplification of error is the result of positive feedback loops between behavior and beliefs, often associated among sociologists with Robert Merton's concept of the "self-fulfilling prophecy." Building on the well-known Thomas theorem, "if men define situations as real, they are real in their consequences" (Thomas 1928, 572), Merton explored the principle of positive feedback in a variety of social contexts. Most notable among them is a fictionalized account of the collapse of a bank following widespread but unfounded rumors of its insolvency, where Merton (1957, 423) notes,

> The parable tells us that public definitions of a situation (prophecies or predictions) become an integral part of the situation and thus affect subsequent developments. This is peculiar to human affairs. It is not found in the world of nature, untouched by human hands. Predictions of the return of Halley's Comet do not influence its orbit. But the rumored insolvency of Millingville's bank did affect the actual outcome. The prophecy of collapse led to its own fulfilment.

Importantly, then, even though both weather forecasting and finance would appear to be consultative in character, in practice, by making predictions about economic behavior, finance ends up *intervening* into economic behavior. The arbitrary or false belief of a relentlessly utility-maximizing *homo economicus* that forms the basis of orthodox economic theory ends up becoming a reality—or so one would have to conclude. The continuing appeal and fascination of the self-fulfilling prophecy lies precisely in its implication that social actors are perpetually caught in a web of their own making (Biggs 2009, 311). Are decision makers judgmental dopes after all?

In recent years, several STS scholars have been reevaluating the thesis of the self-fulfilling prophesy. Instead of the notion that the feedback loops shaping economic behavior are only driven by beliefs and mindsets, they suggest that the "performativity" of finance is fundamentally grounded in the sociotechnical knowledge infrastructures that serve to bring markets into being.[4] In the words of Michel Callon (1998, 2), "economics, in the broad sense of the term, performs, shapes and formats the economy, rather than observes how it functions." By reconceptualizing economics as a material force and by locating the interventionism of economic decision making in its practical, mostly materialized enactment of financial markets, the performativity program analytically allows for events that may well refute or happen independently from what economic actors believe. Thus, even as a series of ethnographic and historical studies of "market devices" (Callon, Millo, and Muniesa 2007; see also Preda 2006)—from the stock ticker and the financial chart to the analyst's report and consumer credit scoring—provide strong empirical evidence for the interventionist-because-representational role of knowledge instruments, other studies identify instances of "counterperformativity" (MacKenzie 2006), where market devices provoke economic behavior that eventually undermines them.

This trend signals what I would describe as a shift from a cognitivist to a pragmatist theory of action. As such, it constitutes a necessary corrective to the decision-making model that underpins the thesis of the self-fulfilling prophecy. The mechanisms driving information feedback loops are now relocated onto discursive and material practices rather than internal belief systems. And the actualization of "wrongheaded" convictions turns into an extreme example of a common social phenomenon— the rousing of a more or less environmentally primed aesthesis into prospective situated action. Fictional expectations about the economy may become real in their consequences, but only if they can be empirically aligned with the sociotechnical context of economic practice. For better or for worse, human reflexivity cannot veer too far or too high in its flights of fancy because it is ontologically anchored in obdurately recalcitrant knowledge infrastructures that keep it in place.

To the extent that they can productively tame nature and make the world calculable, our tools tend to become our world. Hence the performativity of predictions in consultative decision-making fields, such as finance, or in interventionist decision-making fields, such as medicine. Even in meteorology, as we have seen, our devices for capturing and taming the weather, in spite of the fact that it remains untouched by human hands, have grown so elaborate and efficient as to be accepted as a welcome replacement most of the time. That, however, is of little comfort during the heat of the action. The process of decision making in meteorology, medicine, or finance vividly demonstrates that ground truth is a quest, not a conquest. The deep contingencies of the task at hand compel practitioners to improvisational collage in order to craft a locally rational decision. This is as good as it gets for those aspiring to be masters of uncertainty.

Moving Forward

This book has had two main goals. First, it has sought to develop a program for a sociology of decision making. Using a pragmatist theory of action as the backdrop and bringing together scholarship on knowledge production with cognitive, behaviorist, and sociological elaborations of bounded rationality, this program offers an integrated approach to studying decision making in action. Second, this book has sought to demonstrate the utility of this program by fleshing it out through a fine-grained case study of decision-making practice, namely, through a case study of weather forecasting at the NWS. The objective all along has been to allow readers to fully immerse themselves into the rich complexity of weather forecasting practice so that they might come to understand and appreciate for themselves the need for and potential benefits of investigating decision making as practical action in its own right. If nothing else, this book should be read as a call for sociologists to become centrally involved in the study of decision making and for decision-making scholars of all stripes to become centrally committed to a socialized, processual, materialist, and situationist understanding of judgment and decision making.

The question of how far one can convincingly move up the evidentiary chain from a specific case to a general conceptual framework of decision making can only be answered collectively of course, but in this final chapter of the book I have endeavored to make progress toward that goal. Instead of a conclusion, then, I have endeavored to more systematically present my vision for a sociology of decision making in three interrelated ways: first, by schematically articulating how the main analytic components featured in the earlier, empirical chapters—that is, the decision-making environment, the empirical context of action, the art and science of collage, the ends-means practicality of reasoning, the knowledge infrastructure—become entangled in the course of making decisions; second, by theorizing this decision-making model along three analytically distinct dimensions: practice, temporality, and risk; and third, by offering some evidence for its external validity in two other, suitably different, decision-making fields—medicine and finance.

Moving forward, there are already several clear next steps that can be taken to overcome the empirical limitations of this study and broaden its theoretical scope. As already mentioned, I did not get an opportunity to systematically observe the process of expertise acquisition. I elicited a number of fairly rich accounts from Neborough forecasters about how they came to settle on this or that habit of practicing meteorology. I have generally been wary, however, to rely exclusively on ex post facto accounts given my conceptualization of decision making as prospective situated action. Yet an ethnographically-informed longitudinal understanding of the habituation process of decision makers' stock of knowledge is necessary if we are to arrive at a more historicized theory of how institutionalized environment and empirical context are entangled in decision-making practice. As well, my analysis of meteorological decision making and, perhaps more strongly, my discussion of decision making in medicine and finance beg the case for studies that explore the applicability of my argument in fields operating in timescales other than those of meteorology. Given that different temporalities promote different decision-making regimes, new research along these lines promises to fruitfully complicate the schematization of the decision-making process presented in figure 1.

It is to be expected that a sociology of decision making will more readily take roots among sociologists of science and expert organizations. For one, because science and expert organizations contain a considerably higher concentration of externalized decision-making heuristics and techniques and are therefore most amenable to empirical study. For another, because the literature about these fields contains, by extension, a considerably higher concentration of decision-making models and accounts as well as fertile opportunities for collaboration. Yet my call for a sociology of decision making is not confined to STS and organization studies alone. Equally exciting conceptual benefits beckon farther ahead, in the study of social movements, of consumption, of everyday life. Already, a small but growing body of research has drawn renewed attention to the role of prospection in how personal and collective projects, identities, trajectories, and imaginaries arise and are managed.[5] Prospective social action as it actually happens, however, still remains out of grasp. A sociology of decision making represents the next move forward.

Acknowledgments

To use a well-worn but in this case entirely accurate cliché: over the many years that this book has been in the making, I have accumulated many debts of gratitude to many individuals and institutions. The first word of thanks must rightly go to my alma mater, the University of Chicago, where this project began. Andy Abbott saw the promise in my half-hatched idea to study "the negotiations entailed in producing the weather forecast" right from the start and has been an enthusiastic supporter, trusty interlocutor, and honest critic throughout the long process of its metamorphosis into this book. I am fortunate to count him as a mentor and a friend. Andreas Glaeser has deeply influenced how I do and think about cultural sociology. It was Andreas who initially suggested I include the voice of fishers into my analysis, and, although he characteristically continues to insist this was my idea all along, I am delighted that I get to have the final word on the matter. Karin Knorr Cetina joined my dissertation committee, and Chicago, after I returned from the field, but her work has been with me well before this project was conceived. She is a model of generosity and collegiality and, now more than ever, continues to shape the questions I ask about knowledge production. I owe an equally heavy debt of gratitude to Rachel Rinaldo and Robert Wyrod, my fellow dissertators, who back then were always the first to bravely wade through my drafts, and my drafts of drafts. Over the years, I have come to greatly value their camaraderie and counsel—and what a joy it has been to witness our ethnographies see the light of day as books.

If Chicago helped me begin this project, Rutgers helped

ACKNOWLEDGMENTS

me complete it. Ongoing spirited debates with Karen Cerulo, Ann Mische, and Eviatar Zerubavel brought new vitality, clarity, and depth to how I theoretically approach the process of decision making. Many of these fermenting ideas were put to paper during a sabbatical leave in 2011, spent at Princeton's Center for the Study of Social Organization thanks to a gracious invitation by Paul DiMaggio. Viviana Zelizer and the participants of the Workshop on Social Organization offered a healthy dose of conviviality and brilliant conversation during a period of intense writerly isolation. Thanks also to David Gibson for inviting me to the University of Pennsylvania, where I first tested some of the material that appears here.

Several additional colleagues made this book better by offering detailed comments on chapter drafts or related journal articles. I have benefited tremendously from their feedback even if I did not always make use of it. My sincere thanks go to Ulla Berg, Debby Carr, Chip Clarke, Zaire Dinzey-Flores, Gary Fine, Vlad Jankovic, Tom Rudel, Hana Shepherd, and Lucia Timbur. I am especially grateful to Pat Carr, Paul Edwards, Jim Fleming, Judy Gerson, Mike Reay, Audra Walsh, and two anonymous reviewers from the University of Chicago Press for reading the entire manuscript and providing astute suggestions at key stages of its development. Many others helped restore sanity during the writing process with their practical advice and encouragement. I would like particularly to thank József Böröcz, Paul Hirschfield, Lisa Iorillo, Joanna Kempner, Catherine Lee, Julie Phillips, Pat Roos, Zakia Salime, and Arlene Stein for being there when it really mattered.

Close friends and family put up with my obsessive ramblings about this project much longer than was proper or deserved. On this side of the Atlantic, Heather MacIndoe has been a sounding board and confidante through it all. At various occasions, Dan Breslau, Kelly Daley, Rachel Harvey, and Kwai Ng provided culinary and intellectual sustenance when I was hungry for both. On the other side of the Atlantic, Ελένη Βαρουχάκη and Στέλλα Θεοδώρου have been my loudest cheerleaders and my safest havens during the darkest circumstances. I am eternally grateful for the gift that is their friendship. My brother, Έκτωρ Νταϊφάς, takes care of the family front for the both of us—I literally could not be where I am without him. My late mother, Κική Νταϊφά, and my father, Ρήγας Νταϊφάς, gave me every encouragement and opportunity to follow my own path, even if that path would take me so far afield. This book has always been dedicated to them.

I consider myself fortunate to have had my first book manuscript shepherded through production by the legendary Doug Mitchell and his outstanding team at the University of Chicago Press—their professionalism and grace under fire made the process not just tolerable but actually rewarding. Thanks also to Masha Medvedeva, Maria Theresa Patterson, and especially Erica Kees for their invaluable assistance with the interview transcriptions. And thanks to *Poetics* and *Sociological Forum* for publishing earlier or different versions of parts of this book.

My final and most heartfelt words of thanks are reserved for the men and women figuring in these pages: the fishers and broadcast meteorologists who stopped for a moment to talk about the weather with me; and, of course, the NWS meteorologists without whose hospitality and trust none of this would be possible. It has been an honor and a privilege to know and work with them come rain or come shine. Alas, the professional and personal friendships we have developed through the years must remain unacknowledged here for confidentiality reasons. There are, however, three individuals I am able to thank publicly: Jim Stefkovich, then director of the NWS Chicago office, graciously went out of his way to promote my project throughout the agency; Dean Gulezian and Mickey Brown, former director and continuing deputy director, respectively, of Eastern Region Headquarters, gave the final green light on my fieldwork and otherwise generously accommodated my research needs. None of them, nor the NWS, can be assumed to agree with the conclusions of this book. As is customary, I take full responsibility for any errors of fact or interpretation.

Phaedra Daipha
Buenos Aires
August 2014

Notes

INTRODUCTION

1. See Murphy and Winkler (1977), Stewart, Roebber, and Bosart (1997), and Oreskes 2003.
2. For classic formulations of decision making in laboratory science, see Latour and Woolgar (1979), Knorr Cetina (1981), MacKenzie (1981), and Lynch (1985).
3. See, e.g., Simon (1957), March and Simon (1958), Cohen, March, and Olsen (1972), and March and Olsen 1976.
4. See, e.g., Kahneman, Slovic, and Tversky (1982), Gilovich, Griffin, and Kahneman (2002), Rubinstein (1998).
5. The inadequacies and misguided assumptions of Allison's "organizational process" model have been discussed at length by Bendor and Hammond (1992). Recently, Gibson (2012 6–9, 195n65) demonstrated empirically that Allison's model is prone to post hoc fallacies and fails to capture the messiness of decision-making action during the Cuban missile crisis.
6. See, e.g., Perrow (1984), Clarke (1991), and Vaughan (1996).
7. Furthermore, similarly to what Neil Fligstein (1996, 397) has said about organizational neoinstitutionalism, disaster studies tend to have a limited theory of action because they generally focus on how meanings become taken for granted. They are ultimately interested in a sociology of organizations, not a sociology of decision making in its own right.
8. A small number of studies have since attempted to extend Weick's perspective to future-oriented aspects of rhetorical sense making (and sense giving). For a brief review, see Gephart, Topal, and Zhang (2010, 278).
9. For an overview, see Lipshitz et al. (2001).

10. For an insightful effort to draw out the linkages between pragmatism and ethnomethodology, see Emirbayer and Maynard (2011).
11. The list is already too long to be exhaustive but see Joas (1996), Whitford (2002), Sabel (2005), Gross (2009), Mische (2009), Stark (2009), Herrigel (2010), Silver (2011), Berk and Galvan (2009), Schneiderhan (2011), Overdevest (2011), Martin (2011), Lorenzen (2012), Frye (2012), Gibson (2012), and Beckert (2013).
12. Despite claims to the contrary, Simon does not escape this dualist trap, either. His well-known distinction between programmed and nonprogrammed decisions, meant to define two extremes of a continuum, never moves past an either/or elaboration. Ultimately, for Simon (1973), most all problems facing decision makers are ill structured and, therefore, require a nonprogrammed response unless they can be decomposed and serially converted into smaller well-structured problems and programmed decision-making routines.
13. See, e.g., Pliske, Crandall, and Klein (2004), Hoffman and Coffey (2004), Hoffman et al. (2006).
14. In order to as best as possible preserve the anonymity of my informants, I have altered the names and some inconsequential characteristics of the people and places that figure in this book.

CHAPTER 1

1. On doing public science, see Gieryn, Bevins, and Zehr (1985), Zehr (2000), Stilgoe (2007).
2. Reflecting the changing realities and standards of the profession, the American Meteorological Society now requires a college degree in meteorology to grant a "Broadcast Meteorologist" certification. This was not a requirement with its earlier "Seal of Approval" certification program, which was discontinued in 2008.
3. During the eighties, the NWS faced formidable pressure to privatize, and initiatives to upgrade its technology were consistently thwarted. Budget cuts were so severe that the *New York Times* called the NWS a "technological museum" (Mathis 2007, 92).
4. An additional two hundred "satellite" offices essentially functioned as meteorological observatories, taking manual weather observations for one of the main offices.
5. The Automated Surface Observation System was developed in 1993 by the NWS, the Federal Aviation Administration, and the Department of Defense. It comprises a suite of weather sensors that, with several exceptions, measure temperature/dew point, wind speed and direction, pressure, sky condition, basic present weather conditions, visibility, and precipitation accumulation. Data from these sensors are considered "representative" because they are processed through a series of algorithms using a fixed location and time-

averaging techniques. This information is updated once every minute and automatically transmitted to airport traffic controllers, NWS forecasters, and nearby pilots. Currently, there are about a thousand stations in operation throughout the United States. The overwhelming majority have been placed at airports, while the rest are found in mountain passes and remote areas. The Neborough office has 25 such automated weather stations in its area of forecasting responsibility.

6. The National Digital Forecast Database features forecast grids for about twenty land and marine weather elements. Temporal resolution varies but is generally three-hourly for the first three days and six-hourly for the remaining four days of the forecast. Many of these weather elements, however, constitute "derived elements" digitally created from a number of other, more elementary elements that must be worked on at an hourly resolution to be successfully integrated into the Database.

7. Characteristically, in the photo album used by Neborough forecasters to commemorate important events in the office's history, one finds a number of pages dedicated to the installation of the AWIPS technology. There are no pictures commemorating the implementation of the IFPS.

8. Given the male-dominated nature of the field, the personal accounts of female meteorologists are particularly compelling in this respect. Invariably, they include a "turning point," where finally it becomes supremely clear that one must follow the path of weather forecasting despite all odds. In one such story from my interviews, the meteorologist gets accepted to college as an animal science major but spends the summer agonizing over her career decision, when

> Two days before I left for school, I was in a Wendy's with my boyfriend, and there was a thunderstorm in progress, and I kept shushing him so I could count the seconds between lightning and thunder! And I've always been fascinated with the atmosphere, ever since I was very young, I used to go out and measure the snow. If there was a possibility of a snow day, I'd be up at like three o'clock in the morning checking outside: how much are we getting. Thunderstorms, I'd make a bowl of popcorn and just like sit and wait for them to come in. And I would always watch all three meteorologists on the 11 o'clock news, and I'd pit them one against another to see who had the most accurate forecast. So, it was sort of a subtext, a subdialogue, going on in my life that I really didn't identify because there weren't any female role models at that time at all. . . . [That thunderstorm] was a moment of divine intervention. From that moment forward, since that thunderstorm happened, I said "I know exactly what I should do. I should be a meteorologist." I just said, "This is it, I just know it!" And it was a moment, really truly, it was a fated moment, where I just turned the corner and I never ever regretted it, never questioned it again.

9. I was characteristically told about former forecasters who left the Neborough office to take an administrative position at Eastern Region Headquarters that "they sold out and now they have become regionalized," that is, brainwashed to forget their experiences in the field and, presumably, what is really important.
10. While the conventional shorthand for a scientific forecast is its accuracy, the scientific quality of a forecast actually consists of several attributes, including accuracy, reliability, skill, resolution, sharpness, and uncertainty (Stanski, Wilson, and Burrows 1989; Thornes and Stephenson 2001).
11. Per the official NWS weather terminology, "brisk" denotes 15–25 mile per hour winds accompanied by cold temperatures, whereas "breezy" denotes the same wind range but with mild or warm temperatures. The issue for Neborough forecasters, then, was a matter of semantics: Should cold/warm temperatures be understood in absolute or in contextual terms? Is it more meaningful to predict "brisk" winds in the middle of June or should one stick with "breezy" instead?
12. Not surprisingly, the new NWS policy has been built into the IFPS code, which automatically generates a precipitation event in the Weather grid for those points that have a value of 15 percent and above in the Probability of Precipitation grid. Should a forecaster attempt to take the precipitation event out of the Weather grid without changing the Probability of Precipitation grid, she will not be able to publish her forecast to the Digital Database, the IFPS flagging in red the culpable gridpoints.
13. Over the years, a series of small-scale surveys have found that forecast users say they are more likely to change their plans if there is at least a 20 to t30 percent chance of precipitation in the forecast. It is presumably on the strength of these surveys and because this is roughly the daily climatological value for chance of measurable precipitation in much of the eastern United States that 20 percent has conventionally been established as the lowest probability threshold for mentioning precipitation in the forecast.
14. Indeed, a high-ranking NWS official confirmed as much during a private conversation.

CHAPTER 2

1. This is a good place to note that, unlike some other offices, the Neborough office does not feature a weather balloon launch station, traditionally a vaulted structure adjacent to every forecast office. Rather, it is at a Neborough airport that, simultaneously with more than nine hundred such launch stations worldwide, weather balloons are inflated and released twice daily. Radiosonde weather balloons carry an instrument packet that takes measurements of air temperature, humidity, wind speed, and atmospheric pressure to altitudes as high as one hundred thousand feet. These vertical data are transmitted back to the launch station and entered into a worldwide

communications network to be automatically plotted into soundings of the atmosphere.
2. It is this last exit door that is closest to the weather instruments outside, and it is this exit that the staff uses to take the required six-hourly weather observation. However, it is preponderantly the two exits in the operations deck, next to the cubicles and next to the kitchen, that forecasters use for their ad hoc weather observation outings (see chap. 4).
3. This phenomenon is, of course, not unique to the Neborough office or the NWS at large, and has been well studied in STS literature (see, e.g., Shapin 1989, 1994; Barley and Bechky 1994; Orr 1996; Barley and Orr 1997; Doing 2004).
4. Management consists of the Meteorologist-in-Charge, i.e., the office director; the Warning Coordination Manager, in charge of public outreach and hazardous weather preparedness; the Science Operations Officer, in charge of training and research; the Service Hydrologist, in charge of drought and flood risk assessment; the Data Acquisition Manager, in charge of the HMTs and the volunteer weather observer network; and the Electronic Systems Analyst, in charge of the electronics technicians. Office support is provided by the Administrative Support Assistant, the Information Technology Specialist, and three Electronics Technicians.
5. That said, all recent interns at Neborough arrived with a master's degree—weather forecasting has not escaped the rising trend of postgraduate education.
6. As a gendered occupation, weather forecasting rests on a gendered educational discourse and practice. As Berner and Mellstrom (1997, 51) demonstrate for the case of engineering, it requires "a certain kind of man" to be able to master the discipline so as to become an accredited meteorologist. It was, therefore, with unconcealed pride and a gleam in their eye that all the women forecasters I talked to, inside and outside the NWS, would point out and fondly dwell on the fact that when they started out as freshmen in meteorology in the mid eighties to early nineties, "there were two girls and eighty something boys. Now there are many more, but it's still insane. And four years later at graduation, two girls and only four boys!"
7. Not surprisingly, women who had decided not to pursue a forecasting career in the NWS were more forthcoming and reflective about the gendered environment of NWS forecast offices, as the following interview excerpt with a successful broadcast meteorologist, who had interned at Neborough several years ago, illustrates:

> Forecaster: You know, it was such a boys' club . . .
> Phaedra: I suppose it still is to an extent.
> Forecaster: Absolutely, absolutely. I had a great time there. My nickname was "Golden Voice of the Northeast." Because I was the only female on NWS radio. The NWS was an interesting thought but male dominated,

and more than that I wanted to be able to live in a particular part of the country. I wasn't willing to sacrifice that and go anywhere that I was assigned for years and eventually work my way back into the Northeast . . .

Phaedra: So how much of a boys' club was the NWS?

Forecaster: It wasn't that they weren't nice, they were very nice. But I was young, very shy and immature, and it was a very intimidating environment. It took me a long time to find my voice in the science and to feel confident and to feel that people would turn to me and ask me questions or seek out my opinions. And so I just dug deep and worked very hard at it. And when I first came to the Neborough market I remember going to weather conferences and I'd be one of maybe two or three women in a room of 150 people. And if I asked a question, you know, opened my mouth, the whole room would turn and look at me. You know, "Is that a female voice, what is that?" So, that's very intimidating when you're young.

8. It is unclear, and ultimately irrelevant for the purposes of this analysis, why the Neborough office has had such a history of attracting strong personalities. For what it is worth, the standard explanation given by various members of the staff was that the office was bound to attract employees from the broader Neborough area, "and we are known for our crusty temperament."

9. Fine conducted the overwhelming bulk of his fieldwork at the Chicago office, which, as I explain below, provides the closest and most instructive comparison case to the Neborough office.

10. It is thus telling that Fine (2007, 7) only needed to spend ten days each at the Belvedere and Flowerland offices to be able to compare and contrast them to Chicago. To be sure, Fine is a true master at his craft and exceptionally adept at reaching theoretical saturation. Even so, it is thanks to a joint sociotechnical knowledge infrastructure and shared ways of reasoning and doing that local cultural differences among NWS forecast offices can be made so readily recognizable and intelligible.

11. My observations that the workload at Neborough is typical across NWS forecasting offices is further corroborated by a report by the National Research Council (NRC 2012, 53), which acknowledges that forecast offices are "staffed for fair weather."

12. The persisting importance of weather in the wild for the process of meteorological decision making will emerge as a major theme in the next chapter.

13. There is the Model Diagnostic Discussion, which presents model initialization and trend issues and delves into model differences and preferences. There are the 12–48 hours Forecast Discussion, the Extended Forecast Discussion, and the Quantitative Precipitation Discussion. And there is the Marine Weather Discussion. In addition, depending on the time of year and the evolving weather situation, one can choose from the Excessive Rainfall Potential Outlook, the Heavy Snow Discussion, the Mesoscale Discussion, and

the Convective Outlooks from the Storm Prediction Center, or the Tropical Weather Discussion and the Tropical Weather Outlook. Finally, there are the thirty-day and the ninety-day forecast discussions to keep abreast of.
14. To illustrate, neighboring offices would systematically ignore chat messages or provide little communication other than "grids posted" or "[Area Forecast Discussion] in the can." Similarly, they would often wait until the last minute to post their ISC grids, or they would make surprise grid changes that disrupted the previously agreed-on forecast consensus.
15. The conundrum between consensus versus accuracy is encapsulated in the following promoted collaboration practice: "If you have an odd-shaped extension of territory in your [area of forecasting responsibility], and you are the outlier for a given forecast, make a compelling case in the chat room or come into line with your neighbors" (NWS 2003a). On the one hand, then, the case can be made that collaboration actually promotes the science because it gives equal weight to all voices and does not privilege the most overpowering or resistant ones; it promotes meteorological thinking because to go against forecast consensus, a forecaster will have to justify his solution well enough to persuade others. On the other, a well-collaborated forecast is often not the most accurate forecast, but it does tend to prevail in the long run (Howerton 2003).
16. Per NWS policy, two neighboring offices do not have to collaborate if, e.g., differences in their forecast grid for temperature are within five degrees Fahrenheit or within 30 percent for sky cover or 20 percent for probability of precipitation (NWS 2004a). Apparently, these collaboration thresholds have been set high enough to allow for legitimate meteorological differences while weeding out arbitrary discrepancies.
17. After several "embarrassing" exchange episodes during the initial year of the chat room, NWS forecasters were warned by Headquarters, "Don't get cute with it. Someone is always watching."
18. Even so, however, keeping the forecast up-to-date most often does not entail active cognitive engagement or true decision making because it simply involves populating the grids with the latest RUC model run. In fact, keeping the forecast up-to-date is considered so straightforward and mindless a task that it is usually relegated to the HMT or the intern on shift.
19. Text written out from the original shorthand.

CHAPTER 3

1. The biggest challenge is that an automated weather station cannot detect clouds above twelve thousand feet, whereas a human observer has ways of estimating or even measuring cloud cover/visibility. Other issues include reliable measurement of snow, heavy rain, and wind speed.
2. Thus, the Cooperative Weather Observer Program, established in 1890 and comprising a network of approximately eleven thousand citizen volunteers

around the country, mostly entails humans reporting weather measurements by machines sited in their yard into a computer program rather than taking observations themselves. The concern is for more, and more stable, data sites rather than for human sensors per se (for a brief history of the surface weather observation network, see Fiebrich 2009). The original NWS distinction among "complementary," "supplementary," and "augmented" observations, where complementary data originate from automated sources whereas supplementary and augmented data from human observers (the former during routine and the latter during event-driven observations), has long been extinct in practice. In 2009, the separate NWS directives, devoted to complementary and supplementary weather observations, respectively, were combined into one.

3. I consciously opted against Bourdieu's (1990, 57) top-down explanation of "regulated improvisation." Inspired by American pragmatism, my theory of practice posits that decision makers are not regulated by their habits but, instead, that the embodiment of habits is ongoing and intelligent. What I have termed *disciplined improvisation* usually involves, therefore, rule-based and slow—i.e., disciplined—information bricolage and not the seemingly spontaneous and unplanned behavior attendant to Bourdieu's theory.

4. In fact, as reported by Fine (2007, 24), during the so-called modernization of the NWS forecast office network in the early nineties, NWS higher-ups briefly toyed with the idea that forecast offices have no windows so that forecasters would not be distracted by the weather outside. In the eyes of the NWS administration, its forecast offices are "field offices" purely in administrative, not epistemic, terms.

5. In the most recent incident of NWS forecasters successfully mobilizing their social capital to further their professional interests, local and state officials vehemently opposed plans by the financially strapped NWS to centralize its forecasting operations, citing the need for locally knowledgeable and immediately responsive meteorological services.

6. NWS offices are expected to log a "manual weather observation" every six hours. The observation log includes information on sky coverage—percent coverage, cloud type(s), and cloud height(s)—and, when applicable, amount of liquid precipitation.

7. On the fundamentally multisensory processing of the weather outside, see Ingold, 2005. Keeping an ear out does not only pertain to the weather outside, of course. Automated observations of weather events above certain critical thresholds also vie for forecasters' attention by giving off distinctive audio alerts, set to sound every couple of minutes until turned off (on the sounds of the laboratory, see Mody 2005).

8. Indeed, it is the increased sophistication of visualization and simulation technologies that, rather than reinforcing the conventional mind-body split in sociological analysis, has fueled a renewed scholarly interest in body work and the acquisition of expertise (see, e.g., Prentice 2005; Myers 2006).

9. There are obvious connections to the phenomenological tradition here, particularly to Heidegger's analysis of Dasein as "in-der-Welt-sein" but more fundamentally to Merleau-Ponty's perceptualist and synesthetic reading of "être-au-monde" (for a direct extension of Merleau-Ponty's work on the perception of the weather, see Ingold 2005). Meanwhile, my use of "being *there*," as opposed to "being *here*," is also meant to denote that it is still the lab-like environment of the office that houses weather forecasting practice and as such constitutes forecasters' "here."
10. For a psychological argument linking perception with the heuristic of collage, refer to the work of Barbara Tversky (1993), who coined the term *cognitive collage* to suggest an alternative mental model of spatial knowledge to that of the "cognitive map." According to Tversky, the collage metaphor better accounts for the ad hoc multimodal alignment of information during route navigation in the face of the systematic distortions inherent in working memory. Yet, beyond capturing the mentalistic process of perception, the collage metaphor can be fruitfully employed, I argue, to suggest how one goes about actually *crafting* visual coherence.
11. "Les papiers collés dans mes dessins m'ont aussi donné une certitude" (Braque 1917, 5).
12. Note that the minimum criterion for issuing a winter storm warning in the Neborough region has been set to an average of six inches of snow accumulation over a twelve-hour period (see chap. 4).
13. On hypertext as collage, see Landow (2006, 188ff.).

CHAPTER 4

1. Thus, the "weather" grid of the National Digital Forecast Database displays in color code the probability and intensity of, for example, "drizzle," "severe thunderstorms," "snow showers," "sleet," or "fog." But of course even that most circumscribed "weather" is too much weather to be intelligibly portrayed in a single grid and had to be further adjusted. Since November 2005, the Weather grid now displays the "predominant weather" for the given three-hour time period. This operational move centers around the most durable materiality of the weather, doing away with such short-lived meteorological events as "gusty winds," "damaging winds," "hail," "frequent lightning," and "heavy rain" (NWS 2005c). Not that the previous typology included all materialized surface weather—even then, such elusive phenomena as tornadoes did not make the list (NWS 2003b).
2. Off the record, so, too, will senior administrators.
3. Thus, to my knowledge, the subject of their personal verification score has never once come up during forecasters' individual annual performance review meetings with the office director at Neborough. The rationale is prominently featured in the NWS directives: "objectively derived verification scores *by themselves* seldom fully measure the quality of a set of forecasts.

A forecaster demonstrates overall skill through his or her ability to analyze data, interpret guidance, and generate forecasts of maximum utility. Individual forecaster verification data is a private matter between office management and employees and will be safeguarded. To properly utilize forecast verification scores in the performance evaluation process, managers use scores as an indicator of excellence or of need for improvement. For example, a skill score which is "clearly above average" may be used, *in part*, to recognize excellence via the awards system. However, NWS managers at all echelons should be aware that no two forecasters, offices, or management areas face the same series of weather events. Factors which must be taken into account include the number of forecasts produced, availability and quality of guidance, local climatology, and the increased level of difficulty associated with rare events. There is no substitute for sound supervisory judgment in accounting for these influences" (NWS 2009a; emphasis in the original).

4. The fact that, as a rule, NWS automated weather stations are located at airports does not help matters, of course. As one forecaster quipped while discussing NWS verification, "I don't know many people who live, work, or play at an airport."

5. It bears underlining that this double standard of accuracy between non-hazardous and hazardous weather forecasting operations exists beyond the confines of the NWS and guides organizational restructuring recommendations. One thus reads in the latest Congress-commissioned report by the National Research Council (NRC 2012, 55) that in the name of cost effectiveness and efficiency, the NWS might consider, inter alia, regionalizing some of its fair-weather forecasting operations because, "while there may be local weather pattern nuances for each city or county, it is reasonable to think that a team of forecasters with the tools that the NWS provides, including increased [numerical weather prediction] accuracy and associated statistical guidance, would easily be able to produce a forecast that is just as accurate as one produced locally."

6. Unsurprisingly, since the NWS implementation of hazardous weather verification metrics, there has been a well-documented "inflation" of verifying reports, which confounds explanations for the high linear correlation between watches and hazardous weather.

7. A Warning is reserved for conditions posing a direct threat to life or property, while an Advisory is employed for less serious conditions that cause significant inconvenience and that could lead to potentially life- or property-threatening situations if caution is not exercised.

8. In fact, not only has this winter weather event been recorded in Neborough's climatological history as a major snowstorm, but with 38 out of the 38 issued warnings verified, it has also been recorded as a major snowstorm successfully handled (for a discussion of this forecasting event, see Daipha 2007).

9. While forecasting operations at each NWS office follow the rhythm of the local time zone, forecast models' update cycle obeys Coordinated Universal

Time (formerly Greenwich Mean Time), also known as "Z" or "Zulu" time. The 12Z NAM model run, for example, is initialized around 7:00 a.m. and typically arrives at Neborough workstations around 11:00 a.m., just in time for the short-term forecaster to take a quick look at the latest model guidance before starting to do the grids.
10. Press quotations throughout the book have been slightly paraphrased for confidentiality reasons.
11. Had Ray logged a report of three inches and a report of six inches instead, the average snow accumulation for the county would have been four-and-a-half inches, and the winter storm warning would not have been verified.
12. I shadowed the writing up of this case study in its entirety. While there may well have been some discussion privately about the matter, it is worth noting that I witnessed no deliberation whatsoever among the authors as to whether or not to mention the forecast decisions of the midnight shift. It was simply understood that the faux pas in question, while certainly regrettable, was not pertinent information. When I raised the issue with each one of the authors separately, they appeared genuinely mystified by my question, giving me similar explanations as to why it was misguided.
13. To be sure, some organizational performance expectations can become standardized through the implementation of procedural standards and forecasting tools. The technological script of the National Digital Forecast Database and the related NWS directives, for example, have effectively established a performance standard of forecast consistency, as we have already seen.
14. For a recent review of the standardization literature in sociology, see Timmermans and Epstein (2010).

CHAPTER 5

1. This aphorism by Hippocrates is usually rendered in English as

    Life is short,
    [the] art long,
    opportunity fleeting,
    experience misleading,
    judgment difficult.

2. Hence the double entendre in the Hippocratic aphorism, earlier.
3. As a comparison, during a snowstorm, the radar is typically set to scan five angles in ten minutes.
4. See, e.g., Weick and Sutcliffe (2003), Snook (2000), Weick (1993), and Klein, Calderwood, and Clinton-Cirocco (1986).
5. Tellingly, Fine (2007, 45) encounters the same organizational virtue of "keeping cool" during severe weather beyond the realm of NWS forecast offices, at the Storm Prediction Center itself.
6. As a reminder, the minimum hail size criterion for issuing a Severe Thun-

NOTES TO PAGES 156–173

derstorm Warning was three-quarters of an inch (roughly the size of a dime) until 2009. It was increased to one inch (roughly the size of a quarter) to combat desensitization to NWS warnings.

7. To further complicate matters, the attributes of the various dual-process models of cognition cannot be coherently distilled into a single grand theory of dual-processing (Evans 2008). Given the widespread agreement on the basic distinction between a fast, automatic, and implicit system versus a slow, effortful, and explicit system of cognition, however, I will continue to refer to "dual-process theory" for the purposes of this discussion.

8. Recall that, until the NWS digitization of the forecasting routine in 2003, forecasting duties during a typical shift were divided between the land desk and the aviation/marine desk. By 2008, when I returned back to the field, the overwhelming majority of offices had switched over to the short-term/long-term split, with the short-term forecaster also responsible for any aviation/marine duties.

9. To derive this relationship, I privately asked each forecaster to tell me which desk he/she preferred to work at and why. Three forecasters said they preferred working at the short-term desk, two forecasters said they preferred the long-term desk, and the remaining forecasters had no clear preference. At a different occasion, I asked each forecaster to list the three colleagues he/she most enjoyed working with and why. At yet another occasion, I asked each forecaster to list three colleagues he/she thought were good forecasters and why.

CHAPTER 6

1. The 2002 decision by the Weather Channel to discontinue airing the NWS forecast and produce its own local forecasts instead effectively marked the end of NWS's presence on commercial airwaves.
2. See, e.g., Racy (1998), Sink (1995), Krenz and Evans (1993), Murphy and Curtis (1985), and Murphy et al. (1980).
3. Developed in 1994 at the University of Michigan, ACSI employs online random samples to produce a uniform, cross-industry national customer satisfaction index as well as indexes for 10 economic sectors, 47 industries, more than 225 companies, and over 200 federal and local government services.
4. It is conceivable that the Internet and social media will increase opportunities and spaces for dialogic communication between forecasters and the general public. Early signs of a new trend were already evident during my fieldwork. For example, Peter, the Neborough marine focal point, sent out an e-mail in 2008 alerting the staff about "some helpful forecasting rules of thumb for X-Bay we received recently from a local mariner." As I learned later on, Peter found himself in a weather chat room with a recreational boater who shared his thoughts and experiences on X-Bay weather, highlighting failings of NWS forecasts under particular wind regimes. According to Peter,

"the content was solid and consistent with marine forecasting concepts, thus I passed on the info to the staff."
5. Southwest is currently the only commercial carrier to rely exclusively on the NWS for its operations. Unlike other major commercial airlines, which contract out or use their own meteorological staff, Southwest only has two in-house meteorologists, both reserved for extreme situations. This is not to say that Southwest is the only commercial airline to be dependent on NWS forecasts, however. Because the Federal Aviation Administration (FAA) abides by NWS forecasts, so must all airlines as well. For example, if a NWS forecast calls for low visibility, airlines are required by the FAA to carry extra fuel and assign an alternate airport for landing. NWS forecasters, therefore, are quite aware that their aviation forecasts have real and costly consequences attached to them.
6. Since the ill-fated Southwest landing at Chicago's Midway airport during a January snowstorm in 2005, this statement is sadly not true anymore.
7. As an aside, while almanacs (from the Arabic *al-manaakh*, i.e., the climate) remained a full-blown form of folk literature, with notations of anniversaries and curia, home medical advice, statistical miscellany, jokes, and even fiction and poetry (a trope still preserved in the widely known *Old Farmer's Almanac*), it surely is not a coincidence that it was the publication of the *Nautical Almanac* in 1767 that marked the turn to a truly scientific preoccupation with the weather.
8. Reference not provided for confidentiality reasons.
9. Reference not provided for confidentiality reasons.
10. Accordingly, marine forecasts do not include predictions about temperature, a prominent feature of land forecasts, but focus instead on wind, waves, precipitation, visibility, and weather hazards.
11. That the size of the boat one is on would be directly proportional to the amount of wind that boat can withstand seems intuitive enough. Small vessels, which can be up to forty-nine feet long and are considered day boats, typically leave port if the forecast calls for up to fifteen to twenty knot winds. Medium-sized vessels, ranging between fifty to sixty-nine feet and meant for nearshore waters, typically leave port if the forecast calls for up to 20 to 25 knot winds, and they stay at sea for up to five days with a crew of two to three. Large boats measure over seventy feet and can handle upward of forty knot winds. They are primarily rigged with scalloping dredges or otter trawls, and they typically stay offshore for over a week with crews of four to ten.
12. Here is a compelling case in point offered by a gillnetter:

> Phaedra: So would you say you've gone fishing out of greed, even though the forecast was bad?
> Fisher: It's not out of greed in this business with the gillnetting, even though once we get a lot of fish with a high price, then we do get greedy, we want to pile up all we can because right now we're not catching much fish. But what it is with gillnetting is you've got to stay

on the nets, you have to stay on the nets. You have to get them every day, take the fish out because they stay in the water. If they're in the water more than two days, crabs start getting in the nets, seaweed starts getting in the nets, fish start dying, you're going to throw them away. So you have to stay on top of the nets, you've got to keep them clean. Keep them clean, keep them open. If you go take all the fish out, open them back up and put them back, you'll catch more fish that way. If you're fishing for lobsters, you can let the traps stay, they'll catch more lobsters. But with the gillnets, you have to tend them constantly. So the more you tend them, the more fish you get. Like, I don't know if you remember, but this was a real bad January. And the forecast, they'd give like the five-day forecast, they'd call for thirty knots, thirty knots, thirty knots. Every single day! And they were calling for wind right through, so I'm thinking, instead of just waiting it out a week and throwing all the fish away, we might as well go pound around every day and clean them out because it's easier. In real hard wind it's hard to do, it's very hard to keep on the nets. It's very hard for the guys to get the fish out. You can snap the nets, get them caught in the propellers, very hard. So if you get them every day, there's less stuff in there, and it's easier to work it. So when I hear thirty knots, thirty knots, thirty knots, thirty knots, we're just going to keep going, and making it as easy as possible because if we went on a four day set, at thirty knots, we're going to snap the net, we're going to get it stuck, it's going to be full of junk and crabs and snails. So, the more you go, the easier it is. And the more money you'll make.

13. Reference not provided for confidentiality reasons.
14. In Codtown, if a groundfisher cannot find a good fish within the twenty-five miles that comprise the coastal waters, he must venture seventy miles out because the grounds in between are closed to groundfishing.
15. This provision of the old days-at-sea policy was intended to prevent fishers from cheating the system by unloading their catch and returning to the closed area for more fish.
16. The offshore forecast, as opposed to the coastal marine forecast, typically refers to waters beyond twenty-five nautical miles from the coast and is issued by the Ocean Prediction Center in Washington, DC.
17. Because NOAA is the parent organization of the NWS but also of the National Marine Fisheries Service, which fishers refer to as NOAA Fisheries, most fishers said "NOAA" when discussing the NWS with me. The NWS is partially to blame for this misnomer, however. During my fieldwork, the agency started identifying itself as "NOAA's NWS". After two years of complaints from NWS forecasters who felt the agency was losing its identity ("selling out" was one of the phrases used), this continuing trend has been more subtle.
18. In the Neborough area and adjacent offshore waters, for example, there are

only eight such buoy stations, of which one or two are usually inoperative because of extreme weather conditions or vandalism by (inebriated) fishers.
19. It is for the same reason that the NWS has not been successful in enlisting fishers as marine weather spotters.
20. Indeed, it would seem that Codtown and Whaletown commercial fishers have developed a veritable joke repertoire about NWS forecasting, which they were quite happy to rehearse for my benefit:

> The NOAA weather office is pretty good. Except when they screw up big time. [*laughter*] Yeah, all the fishermen joke about it. All the fishermen. Because when they say ten to twenty diminishing, we turn that into "add them up and demolishing." [*laughter*] Because sometimes they'll say it's going to be windy in the morning but diminish in the afternoon. So we'll say, all right, we'll put the rough morning in, then it'll get nice. And it never does, it gets worse, and we notice that a lot. All the fishermen joke about the weather being wrong. Constantly. The only thing that's really good for us is the buoy reports. It doesn't lie. It tells you what the wind's blowing at what hour. And you can get that on the computer, too.

* * *

> You know, there are jokes around here. "What's the wind going to be? Variable to terrible." [*laughter*] That's like shooting craps in Atlantic City, you know? You don't know what you're going to come up with until after it rolls. I mean, we'd be out there, it'd be blowing thirty, and they're saying variable ten knots or less. Well, we're like "What the hell? In their office maybe it's ten knots or less." [*laughter*]

21. This propensity has occasionally led the NWS to unfortunate extremes, such as the slogan "America's no surprise Weather Service" that defined the NWS 2000–2005 Strategic Plan, much to the chagrin of NWS forecasters and the glee of media.
22. More specifically, the survey yielded a 98% response rate with a total of 140 respondents, of which 110 were recreational boaters.
23. For recent research on the social effects of weather information, see, e.g., Lindell and Brooks (2013), Morss, Lazo, and Demuth (2010), Lazo, Morss, and Demuth (2009), NRC (2006), Stewart, Pielke, and Nath (2004). For an authoritative review of natural disasters scholarship, see Mileti (1999).
24. See, among others, NRC (2006), Durante et al. (2005), Grimit and Mass (2004), Zhu et al. (2002), Katz and Murphy 1997.

CHAPTER 7

1. This argument is not far from Swidler's (1986, 2001, 100) point that cultural repertoires have a "greater, or at least more obvious" and explicit influence on action during "unsettled" versus "settled" periods. After all, unsettled

circumstances are *premised* on an empirical context that does not seem to support what were erstwhile typical recipes for action. It is because we first empirically realize we are not capable of handling a situation that we *then* reach out for alternative solutions to and explanations for the problem at hand. Even in periods of great cultural change, Swidler (1986, 103) notes that "an ideology's appeal may depend on the strategies of action it supports; and the specific historical situations . . . may determine which [cultural models] take root and thrive and which wither and die." What my proposed framework does contribute to the cultural analysis of change is a potential way out of the admittedly rigid dichotomy between settled and unsettled fields of action.

2. Note, therefore, that the extent to which decision makers can rely on their cultural repertoires to handle a fateful situation depends on the extent of institutionalization of the decision-making environment in which they operate but, crucially, also on the extent of their empirical exposure to similar situations. Expertise acquisition is self-energized and context dependent.

3. See, e.g., Bechky (2003), Carlile (2004), and Gorman (2010).

4. See, e.g., Callon (1998), MacKenzie (2006), and MacKenzie, Muniesa, and Sieu (2007).

5. For theoretically programmatic contributions, see Mische (2009), Tavory and Eliasoph (2013).

References Cited

Abbott, Andrew. 1981. "Status and Status Strain in the Professions." *American Journal of Sociology* 86: 819–35.
———. 2001a. *Chaos of Disciplines*. Chicago: University of Chicago Press.
———. 2001b. *Time Matters: On Theory and Method*. Chicago. University of Chicago Press.
———. 2004. *Methods of Discovery*. New York: Norton.
Abolafia, Mitchel Y. 1996. *Making Markets: Opportunism and Restraint on Wall Street*. Cambridge, MA: Harvard University Press.
Akrich, Madeleine. 1992. "The De-Scription of Technical Objects." In *Shaping Technology, Building Society: Studies in Socio-technical Change*, edited by Wiebe Bijker and John Law, 205–24. Cambridge, MA: MIT Press.
Allison, Graham, and Philip Zelikow. 1999. *Essence of Decision: Explaining the Cuban Missile Crisis*. 2nd ed. New York: Longman.
Anderson, Katharine. 2005. *Predicting the Weather: Victorians and the Science of Meteorology*. Chicago: University of Chicago Press.
Armstrong, Elizabeth M. 2003. *Conceiving Risk, Bearing Responsibility: Fetal Alcohol Syndrome and the Diagnosis of Moral Disorder*. Baltimore: Johns Hopkins University Press.
Atkinson, Paul. 1984. "Training for Certainty." *Social Science and Medicine* 19: 949–56.
Bakken, Tore, and Tor Hernes. 2006. "Organizing Is Both a Verb and a Noun: Weick Meets Whitehead." *Organization Studies* 27: 1599–616.
Barley, Stephen R., and Beth A. Bechky. 1994. "In the Backrooms of Science: The Work of Technicians in Science Labs." *Work and Occupations* 21: 85–126.

REFERENCES CITED

Barley, Stephen R., and Julian E. Orr, eds. 1997. *Between Craft and Science: Technical Work in US Settings*. Ithaca, NY: Cornell University Press.

Batten, Frank, and Jeffrey L. Cruikshank. 2002. *The Weather Channel: The Improbable Rise of a Media Phenomenon*. Boston: Harvard Business School Press.

Bechky, Beth A. 2003. "Sharing Meaning across Occupational Communities: The Transformation of Understanding on a Production Floor." *Organization Science* 14: 312–30.

Beckert, Jens. 1996. "What Is Sociological about Economic Sociology? Uncertainty and the Embeddedness of Economic Action." *Theory and Society* 25: 803–40.

———. 2013. "Imagined Futures: Fictional Expectations in the Economy." *Theory and Society* 42: 219–40.

Bendor, Jonathan, and Thomas H. Hammond. 1992. "Rethinking Allison's Models." *American Political Science Review* 86: 301–22.

Berg, Marc. 1997. *Rationalizing Medical Work: Decision Support Techniques and Medical Practices*. Cambridge, MA: MIT Press.

Bergson, Henri. 1946. *The Creative Mind: An Introduction to Metaphysics*. New York: Philosophical Library.

Berk, Gerald, and Dennis Galvan. 2009. "How People Experience and Change Institutions: A Field Guide to Creative Syncretism." *Theory and Society* 38: 543–80.

Berner, Boel, and Ulf Mellstrom. 1997. "Looking for Mister Engineer: Understanding Masculinity and Technology at Two Fin de Siècles". In *Gendered Practices: Feminist Studies of Technology and Society*, edited by Boel Berner, 39–68. Stockholm: Almqvist and Wiksell.

Beunza, Daniel, Iain Hardie, and Donald MacKenzie. 2006. "A Price Is a Social Thing: Towards a Material Sociology of Arbitrage." *Organization Studies* 27: 721–45.

Beunza, Daniel, and David Stark. 2004. "Tools of the Trade: The Socio-Technology of Arbitrage in a Wall Street Trading Room." *Industrial and Corporate Change* 13: 369–400.

———. 2012. "From Dissonance to Resonance: Cognitive Interdependence in Quantitative Finance." *Economy and Society* 41: 383–417.

Biggs, Michael. 2009. "Self-Fulfilling Prophecies." In *The Oxford Handbook of Analytical Sociology*, edited by Peter Hedström and Peter Bearman, 294–314. Oxford: Oxford University Press.

Bourdieu, Pierre. 1977. *Outline of a Theory of Practice*. Cambridge: Cambridge University Press.

———. 1990. *The Logic of Practice*. Stanford, CA: Stanford University Press.

Bowker, Geoffrey C. 1994. *Science on the Run: Information Management and Industrial Geophysics at Schlumberger, 1920–1940*. Cambridge, MA: MIT Press

Bowker, Geoffrey C., and Susan L. Star. 1999. *Sorting Things Out: Classification and Its Consequences*. Cambridge, MA: MIT Press.

Braque, Georges. 1917. "Pensées et Réflexions sur la Peinture." *Nord-Sud* 10: 3–5.

Brekhus, Wayne. 1998. "A Sociology of the Unmarked: Redirecting Our Focus." *Sociological Theory* 16: 34–51.
Burri, Regula V. 2008. "Doing Distinctions: Boundary Work and Symbolic Capital in Radiology." *Social Studies of Science* 38: 35–62.
Burt, Ronald S. 1995. *Structural Holes: The Social Structure of Competition*. Cambridge, MA: Harvard University Press.
Bush, Lawrence. 2011. *Standards: Recipes for Reality*. Cambridge, MA: MIT Press.
Callon, Michel. 1986. "Some Elements of a Sociology of Translation: Domestication of the Scallops and the Fishermen of St Brieuc Bay." In *Power, Action and Belief: A New Sociology of Knowledge*, edited by John Law, 196–233. London: Routledge and Kegan Paul.
———. 1998. *The Laws of the Markets*. Oxford: Blackwell.
Callon, Michel, Yuval Millo, and Fabian Muniesa. 2007. *Market Devices*. Oxford: Blackwell.
Calvert, Edgar B. 1899. "Development of the Daily Weather Map." *U.S. Weather Bureau Bulletin* 24: 144–50.
Carlile, Paul R. 2004. "Transferring, Translating, and Transforming: An Integrative Framework for Managing Knowledge across Boundaries." *Organization Science* 15: 555–68.
Caronna, C. 2011. "Clash of Logics, Crisis of Trust: Entering the Era of Public For-Profit Health Care?" In *The Handbook of Health, Illness, and Healing: Blueprint for the 21st Century*, edited by B. Pescosolido, J. Martin, J. McLeod, and A. Rogers, 255–71. New York: Springer.
Cassell, Eric J. 1985. *Talking with Patients: Volume 2, Clinical Technique*. MIT Press Series on the Humanistic and Social Dimensions of Medicine 2. Cambridge, MA: MIT Press.
———. 1991. *The Nature of Suffering and the Goals of Medicine*. Oxford: Oxford University Press.
Chambliss, Daniel F. 1996. *Beyond Caring: Hospitals, Nurses, and the Social Organization of Ethics*. Chicago: University of Chicago Press.
Christakis, Nicholas. 1997. "The Ellipsis of Prognosis in Modern Medical Thought." *Social Science and Medicine* 44: 301–15.
———. 1999. *Death Foretold: Prophesy and Prognosis in Medical Care*. Chicago: University of Chicago Press.
Clark, Andy. 1998. *Being There: Putting Brain, Body, and World Together Again*. Cambridge, MA: MIT Press.
———. 2003. *Natural Born Cyborgs: Minds, Technologies, and the Future of Human Intelligence*. Oxford: Oxford University Press.
Clark, Andy, and David J. Chalmers. 1998. "The Extended Mind." *Analysis* 58: 7–19.
Clarke, Adele E., and Joan H. Fujimura. 1992. *The Right Tools for the Job: At Work in Twentieth-Century Life Sciences*. Princeton, NJ: Princeton University Press.
Clarke, Lee. 1991. *Acceptable Risk? Making Decisions in a Toxic Environment*. Berkeley: University of California Press.

REFERENCES CITED

Clemens, Elisabeth S., and James M. Cook. 1999. "Politics and Institutionalism: Explaining Durability and Change." *Annual Review of Sociology* 25: 441–66.

Cloutier, Charlotte, and Ann Langley. 2013. "The Logic of Institutional Logics: Insights from French Pragmatist Sociology." *Journal of Management Inquiry* 22: 360–80.

Cohen, Michael D., James G. March, and Johann P. Olsen. 1972. "A Garbage Can Model of Organizational Choice." *Administrative Science Quarterly* 17: 1–25.

Collins, Harry, and Robert Evans. 2002. "The Third Wave of Science Studies: Studies of Expertise and Experience." *Social Studies of Science* 32: 235–96.

———. 2006. *Rethinking Expertise*. Chicago: University of Chicago Press.

Colman, Brad. 1997. "'What Is a Good Forecast?': In the Eyes of a Forecaster." Presented at the Workshop on the Social and Economic Impacts of Weather, Boulder, Colorado, April 2–4. http://sciencepolicy.colorado.edu/socasp/weather1/colman.html.

Cox, John D. 2002. *Storm Watchers: The Turbulent History of Weather Prediction from Franklin's Kite to El Niño*. Hoboken, NJ: Wiley.

Craft, Erik D. 1999. "Private Weather Organizations and the Founding of the United States Weather Bureau." *Journal of Economic History* 59: 1063–71.

Crane, Diana. 1975. *The Sanctity of Social Life: Physician's Treatment of Critically Ill Patients*. New York: Russell Sage Foundation.

Csikszentmihalyi, Mihaly. 1990. *Flow: The Psychology of Optimal Experience*. New York: Harper and Row.

Daipha, Phaedra. 2007. "Masters of Uncertainty: Weather Forecasters and the Quest for Ground Truth." PhD diss., University of Chicago.

———. 2010. "Visual Perception at Work: Lessons from the World of Meteorology." *Poetics* 38: 150–64.

———. 2013. "Screenwork as the Social Organization of Expertise." Paper presented at the American Sociological Association Meetings, New York, August 10–13.

———. 2015a. "From Bricolage to Collage: The Making of Decisions at a Weather Forecast Office." *Sociological Forum* 30 (3). Forthcoming.

———. 2015b. "The Role of Prognosis in Cardiology Practice." Paper presented at the American Sociological Association Meetings, Chicago, August 22–25.

Daston, Lorraine, and Peter Galison. 2007. *Objectivity*. New York: Zone.

De Keyser, Véronique, and Anne-Sophie Nyssen. 2001. "The Management of Temporal Constraints in Naturalistic Decision Making: The Case of Anesthesia." In *Linking Expertise and Naturalistic Decision-Making*, edited by Eduardo Salas and Gary Klein, 171–188. Mahwah, NJ: Lawrence Erlbaum.

Dequech, David. 2006. "The New Institutional Economics and the Behaviour under Uncertainty." *Journal of Economic Behavior and Organization* 59: 109–31.

Desmond, Matthew. 2007. *On the Fireline: Living and Dying with Wildland Firefighters*. Chicago: Chicago University Press.

Dewey, John. 1910. *How We Think*. Lexington: D.C. Heath.

———. 1922. *Human Nature and Conduct: An Introduction to Social Psychology*. New York: Holt.

———. 1925. *Experience and Nature.* Chicago: Open Court.
———. 1939. *Theory of Valuation.* Chicago: University of Chicago Press.
DiMaggio, Paul. 1997. "Culture and Cognition." *American Review of Sociology* 23: 263–87.
DiMaggio, Paul, and Walter W. Powell. 1983. "The Iron Cage Revisited: Institutional Isomorpohism and Collective Rationality in Organizational Fields." *American Sociological Review* 48: 147–60.
Doing, Park. 2004. "Lab Hands and the 'Scarlet O': Epistemic Politics and (Scientific) Labor." *Social Studies of Science* 34: 299–23.
Doswell, Charles A. 1990. "Comments on the Need for Augmentation in Automated Surface Observations." *National Weather Digest* 15: 29–30.
———. 2004. "Weather Forecasting by Humans: Heuristics and Decision-making." *Weather Forecasting* 19: 1115–26.
Dreyfus, Hubert L. 1992. *What Computers Still Can't Do: A Critique of Artificial Reason.* Cambridge, MA: MIT Press.
Dreyfus, Hubert L., and Stuart E. Dreyfus. 2008. "Beyond Expertise: Some Preliminary Thoughts on Mastery." In *A Qualitative Stance: Essays in Honor of Steiner Kvale*, edited by Klaus Nielsen et al., 113–24. Aarhus: Aarhus University Press.
Durante, Andrew, Robert Hart, Irv Watson, Richard Grumm, and Walter Drag. 2005. "The Development of Forecast Confidence Measures Using NCEP Ensembles and Their Real-Time Implementation within NWS Web-Based Graphical Forecasts." Paper presented at the 21st American Meteorological Society Meeting, Washington, DC.
Edwards, Paul N. 2006. "Meteorology as Infrastructural Globalism." *Osiris* 21: 229–50.
———. 2010. *A Vast Machine: Computer Models, Climate Data, and the Politics of Global Warming.* Cambridge, MA: MIT Press.
Edwards, Paul N., Steven J. Jackson, Melissa K. Chalmers, Geoffrey C. Bowker, Christine L. Borgman, David Ribes, Matt Burton, and Scout Calvert. 2013. *Knowledge Infrastructures: Intellectual Frameworks and Research Challenges.* Report of a workshop sponsored by the National Science Foundation and the Sloan Foundation, University of Michigan School of Information, May 25–28, 2012.
Emirbayer, Mustafa, and Ann Mische. 1998. "What is Agency?" *American Journal of Sociology* 103: 962–1023.
Emirbayer, Mustafa, and Douglas Maynard. 2011. "Pragmatism and Ethnomethodology." *Qualitative Sociology* 34: 221–61.
Evans, Jonathan St. B. T. 2008. "Dual-Processing Accounts of Reasoning, Judgment and Social Cognition." *Annual Review of Psychology* 59: 255–78.
———. 2010. *Thinking Twice: Two Minds in One Brain.* Oxford: Oxford University Press.
———. 2012. "Questions and Challenges for the New Psychology of Reasoning." *Thinking and Reasoning* 18: 5–31.

REFERENCES CITED

Faraj, Samer, and Yan Xiao. 2006. "Coordination in Fast-Response Organizations." *Management Science* 52: 1155–69.

Federal Signal. 2012. "Revealing Americans' Awareness and Preparedness Surrounding Emergency Situations." http://www.alertnotification.com/pdf/ANS104_2012_Survey-lowRes.pdf.

Fiebrich, Christopher, A. 2009. "The History of Surface Weather Observations in the United States." *Earth-Science Reviews* 93: 77–84.

Fine, Gary A. 1979. "Small Groups and Culture Creation: The Idioculture of Little League Baseball Teams." *American Sociological Review* 44: 733–45.

———. 2006. "Shopfloor Cultures: The Idioculture of Production in Operational Meteorology." *Sociological Quarterly* 47: 1–19.

———. 2007. *Authors of the Storm: Meteorologists and the Culture of Prediction*. Chicago: University of Chicago Press.

Fine, Gary A., and Tim Hallett. 2014. "Group Cultures and the Everyday Life of Organizations: Interaction Orders and Meso-Analysis." *Organization Studies* 35: 1773–92.

Fishman, Mark. 1982. "News and Nonevents: Making the Visible Invisible." In *Individuals in Mass Media Organizations: Creativity and Constraint*, edited by J. S. Ettema and D. C. Whitney, 219–40. Thousand Oaks, CA: Sage.

Fleck, Ludwik. 1979. *The Genesis and Development of a Scientific Fact*. Chicago: University of Chicago Press.

Fleming, James R. 1990. *Meteorology in America, 1800–1870*. Baltimore: Johns Hopkins University Press.

———. 2012. *Fixing the Sky: The Checkered History of Weather and Climate Control*. New York: Columbia University Press.

Fligstein, Neil. 1996. "Social Skill and Institutional Theory." *American Behavioral Scientist* 40: 397–405.

Forgan, Sophie. 1994. "The Architecture of Display: Museums, Universities and Objects in Nineteenth-Century Britain." *History of Science* 32: 139–62.

Fort, Tom. 2006. *Under the Weather*. London: Arrow.

Friedland, Roger, and Robert R. Alford. 1991. "Bringing Society Back In: Symbols, Practices, and Institutional Contradictions." *New Institutionalism in Organization Analysis*, edited by Walter W. Powell and Paul J. DiMaggio, 232–66. Chicago: University of Chicago Press.

Friedman, Robert Marc. 1989. *Appropriating the Weather: Vilhelm Bjerknes and the Construction of a Modern Meteorology*. Ithaca, NY: Cornell University Press.

Frye, Margaret. 2012. "Bright Futures in Malawi's New Dawn: Educational Aspirations as Assertions of Identity." *American Journal of Sociology* 117: 1565–624.

Garfinkel, Harold. 1967. *Studies in Ethnomethodology*. Englewood Cliffs, NJ: Prentice-Hall.

Gasser, Les. 1986. "The Integration of Computing and Routine Work." *ACM Transactions on Office Information Systems* 4: 257–70.

Gazelius, Stig. 2007. "Can Norms Account for Strategic Action? Information Management in Fishing as a Game of Legitimate Strategy." *Sociology* 41: 201–18.

Gephart, Robert P., Topal Cagri, and Zhen Zhang. 2010. "Future Oriented Sensemaking: Temporalities and Institutional Legitimation." In *Process, Sensemaking, and Organizing*, edited by Tor Hernes and Sally Maitlis, 275–303. Oxford: Oxford University Press.

Gerson, Elihu M., and Susan Leigh Star. 1986. "Analyzing Due Process in the Workplace." *ACM Transactions on Office Information Systems* 4: 257–70.

Gibson, David R. 2012. *Talk at the Brink: Deliberation and Decision during the Cuban Missile Crisis*. Princeton, NJ: Princeton University Press.

Giddens, Anthony. 1984. *The Constitution of Society: Outline of the Theory of Structuration*. Cambridge: Polity.

———. 1991. *Modernity and Self Identity: Self and Society in the Late Modern Age*. Stanford, CA: Stanford University Press.

Gieryn, Thomas F. 1999. *Cultural Boundaries of Science: Credibility on the Line*. Chicago: University of Chicago Press.

———. 2000. "A Space for Place in Sociology." *Annual Review of Sociology* 26: 463–96.

Gieryn, Thomas F., George M. Bevins, and Stephen C. Zehr. 1985. "Professionalization of American Scientists: Public Science in the Creation/Evolution Trials." *American Sociology Review* 50: 392–409.

Gigerenzer, Gerd. 2004. "Fast and Frugal Heuristics: The Tools of Bounded Rationality." In *Blackwell Handbook of Judgment and Decision-Making*, edited by Derek J. Kohler and Nigel Harvey, 62–88. Oxford: Blackwell.

———. 2007. *Gut Feeling: The Intelligence of the Unconscious*. London: Penguin.

Gigerenzer, Gerd, Peter M. Todd, and the ABC Research Group. 1999. *Simple Heuristics That Make Us Smart*. New York: Oxford University Press.

Gigerenzer, Gerd, and Reinhard Selten. 2002. *Bounded Rationality: The Adaptive Toolbox*. Cambridge, MA: MIT Press.

Gilovich, Thomas D., Dale Griffin, and Daniel Kahneman. 2002. *Heuristics and Biases: The Psychology of Intuitive Judgment*. Cambridge: Cambridge University Press.

Goffman, Erving. 1959. *The Presentation of Self in Everyday Life*. New York: Anchor.

———. 1967. "Where the Action Is." In *Interaction Ritual: Essays on Face-to Face Behavior*, 149–270. New York: Doubleday Anchor.

———. 1971. *Relations in Public: Microstudies of the Public Order*. New York: Basic.

———. 1981. *Forms of Talk*. Oxford: Blackwell.

Golinski, Jan. 2007. *British Weather and the Climate of Enlightenment*. Chicago: Chicago University Press.

Goodwin, Charles. 1994. "Professional Vision." *American Anthropologist* 96: 606–33.

Gorman, Michael E. 2010. *Trading Zones and Interactional Expertise: Creating New Kinds of Collaboration*. Cambridge, MA: MIT Press.

Grimit, Eric P., and Clifford F. Mass. 2004. "Forecasting Mesoscale Uncertainty: Short-Range Ensemble Forecast Error Predictability." Paper presented at the 16th Conference on Numerical Weather Prediction, Seattle, WA, January 10–12.

Groopman, Jerome. 2007. *How Doctors Think*. New York: Houghton Mifflin.

REFERENCES CITED

Gross, Alan G. 1994. "The Roles of Rhetoric in the Public Understanding of Science." *Public Understanding of Science* 3: 3–23.

Gross, Neil. 2009. "A Pragmatist Theory of Social Mechanisms." *American Sociological Review* 74: 358–79.

Hallett, Tim, and Marc Ventresca. 2006. "Inhabited Institutions: Social Interactions and Organizational Forms in Gouldner's *Patterns of Industrial Bureaucracy*." *Theory and Society* 35: 213–36.

Hammond, Kenneth R. 1996. *Human Judgment and Social Policy: Irreducible Uncertainty, Inevitable Error, Unavoidable Injustice*. Oxford: Oxford University Press.

Hardie, Iain, and Donald MacKenzie. 2007. "Constructing the Market Frame: Distributed Cognition and Distributed Framing in Financial Markets." *New Political Economy* 12: 389–403.

Harper, Kristine C. 2003. "Research from the Boundary Layer: Civilian Leadership, Military Funding and the Development of Numerical Weather Prediction (1946–55)." *Social Studies of Science* 33: 667–96.

———. 2008. *Weather by the Numbers: The Genesis of Modern Meteorology*. Cambridge, MA: MIT Press.

Harris Interactive. 2007. "Local Television News Is the Place for Weather Forecasts for a Plurality of Americans." *Harris Poll* no. 118, November 28, 2007. http://www.harrisinteractive.com/harris_poll/index.asp?PID=839.

Herrigel, Gary. 2010. *Manufacturing Possibilities: Creative Action and Industrial Recomposition in the US, Germany, and Japan*. Oxford: Oxford University Press.

Hickman, Larry A. 1990. *John Dewey's Pragmatic Technology*. Bloomington: Indiana University Press.

Hilgartner, Stephen. 1990. "The Dominant View of Popularization: Conceptual Problems, Political Uses." *Social Studies of Science* 20: 519–39.

Ho, Karen. 2009. *Liquidated: An Ethnography of Wall Street*. Durham, NC: Duke University Press.

Hochschild, Arlie, R. 1983. *The Managed Heart: Commercialization of Human Feeling*. Berkeley: University of California Press.

Hoffman, Robert R., and John W. Coffey. 2004. "Weather Forecasting and the Principles of Complex Cognitive Systems." In *Proceedings of the 48th Meeting of the Human Factors and Ergonomics Society*, 315–19. Santa Monica, CA: Human Factors and Ergonomics Society.

Hoffman, Robert R., John W. Coffey, Kenneth M. Ford, and Joseph D. Novak. 2006. "A Method for Eliciting, Preserving, and Sharing the Knowledge of Forecasters." *Weather and Forecasting* 21: 416–28.

Hojat, Mohammadreza, Michael J. Vergare, Kaye Maxwell, George Brainard, Steven K. Herrine, Gerald A. Isenberg, Jon Veloski, and Joseph S. Gonnella. 2009. "The Devil Is in the Third Year: A Longitudinal Study of Erosion of Empathy in Medical School. *Academic Medicine* 84:1182–91.

Hooke, William H., and Roger A. Pielke Jr. 2000. "Short-Term Weather Prediction: An Orchestra in Search of a Conductor." In *Prediction: Science, Decision-*

Making and the Future of Nature, edited by Daniel Sarewitz, Roger A. Pielke Jr., and Radford Byerly Jr., 61–83. Washington, DC: Island Press.

Howerton, Paul. 2003. "Collaboration in the IFPS Era." http://www-md.fsl.noaa.gov/IFPS/mws4/Collaboration_bird.ppt (no longer posted).

Howerton, Paul, and Shannon White. 2003. "National Collaboration Training Session." http://www-md.fsl.noaa.gov/IFPS/collabTalkPts.txt (no longer posted).

Hughes, Everett C. 1958. *Men and Their Work*. Glencoe, IL: Free Press.

Hughes, Patrick. 1970. *A Century of Weather Service: A History of the Birth and Growth of the NWS*. New York: Gordon and Breach.

Hutchins, Edwin. 1995a. *Cognition in the Wild*. Cambridge, MA: MIT Press.

———. 1995b. "How a Cockpit Remembers Its Speeds." *Cognitive Science* 19: 265–88.

Ingold, Tim. 2005. "The Eye of the Storm: Visual Perception and the Weather." *Visual Studies* 20: 97–104.

———. 2007. "Earth, Sky, Wind, and Weather." *Journal of the Royal Anthropological Institute* 13: 19–38.

Jackson, Myles. 1999. "Illuminating the Opacity of Achromatic Lens Production: Joseph Fraunhofer's Use of Monastic Architecture and Space as a Laboratory." In *Architecture and Science*, edited by Peter Galison and Emily Thompson, 423–55. Cambridge, MA: MIT Press.

James, William. 1907. *Pragmatism. A New Name for Some Old Ways of Thinking*. New York: Longmans Green.

———. 1909. *The Meaning of Truth: A Sequel to "Pragmatism."* New York: Longmans Green.

Janis, Harriet, and Rudi Blesh. 1967. *Collage: Personalities, Concepts, Techniques*. Philadelphia: Chilton.

Janković, Vladimir. 2000. *Reading the Skies: A Cultural History of English Weather, 1650–1820*. Chicago: University of Chicago Press.

Jasper, James. 2012. "Feeling-Thinking Processes: Beyond the Idealism of Cultural Sociology." Lecture presented at the Rutgers Sociology Colloquium Series.

Jenkins Report. 2000. "House of Lords, Science and Technology Committee: Third Report." http://www.parliament.the-stationery-office.co.uk/pa/ld199900/ldselect/ldsctech/38/3801.htm.

Joas, Hans. 1996. *The Creativity of Action*. Chicago: University of Chicago Press.

Kahneman, Daniel. 2011. *Thinking, Fast and Slow*. New York: Farrar, Straus and Giroux.

Kahneman, Daniel, and Gary Klein. 2009. "Conditions for Intuitive Expertise: A Failure to Disagree." *American Psychologist* 64: 515–26.

Kahneman, Daniel, Paul Slovic, and Amos Tversky. 1982. *Judgment and Uncertainty: Heuristics and Biases*. Cambridge: Cambridge University Press.

Katz, Jack. 1997. "Ethnography's Warrants." *Sociological Methods and Research* 25: 391–413.

Katz, Richard W., and Allan H. Murphy. 1997. *Economic Value of Weather and Climate Forecasts*. Cambridge: Cambridge University Press.

Kellogg, Katherine C., Wanda J. Orlikowski, and JoAnne Yates. 2006. "Life in the Trading Zone: Structuring Coordination across Boundaries in Postbureaucratic Organizations." *Organization Science* 17: 22–44.

Klein, Gary A. 1999. *Sources of Power: How People Make Decisions.* Cambridge, MA: MIT Press.

Klein, Gary A., R. Calderwood, and A. Clinton-Cirocco. 1986. "Rapid Decision-Making on the Fireground." In *Proceedings of the 30th Annual Human Factors Society*, vol. 1, 576–80. Santa Monica, CA: Human Factors Society.

Klein, Gary A., Brian Moon, and Robert R. Hoffman. 2006. "Making Sense of Sensemaking II: A Macrocognitive Model." *IEEE Intelligent Systems* 21: 88–92.

Klein, Roberta, and Roger A. Pielke Jr. 2002. "Bad Weather? Then Sue the Weatherman! Part I. Legal Liability for Public Sector Forecasts." *Bulletin of the American Meteorological Society* 83: 1791–99.

Knorr Cetina, Karin. 1981. *The Manufacture of Knowledge: An Essay on the Constructivist and Contextual Nature of Science.* New York: Pergamon.

———. 1997. "Sociality with Objects: Social Relations in Postsocial Knowledge Societies." *Theory, Culture, and Society* 14: 1–30.

———. 1999. *Epistemic Cultures: How the Sciences Make Knowledge.* Cambridge, MA: Harvard University Press.

———. 2001. "Objectual Practice." In *The Practice Turn in Contemporary Theory*, edited by Theodor R. Schatzki, Karin Knorr Cetina, and Eike von Savigny, 175–88. London: Routledge.

———. 2003. "From Pipes to Scopes: The Flow Architecture of Financial Markets." *Distinction* 7: 7–23.

Knorr Cetina, Karin, and Klaus Amann. 1990. "Image Dissection in Natural Scientific Inquiry." *Science, Technology, and Human Values* 15: 259–83.

Knorr Cetina, Karin, and Urs Bruegger. 2002. "Traders' Engagement with Markets: A Postsocial Relationship." *Theory, Culture and Society* 19: 161–85.

Kohler, Robert E. 2002. *Landscapes and Labscapes: Exploring the Lab-Field Border in Biology.* Chicago: University of Chicago Press.

Krenz, S. H., and J. S. Evans. 1993. "Weather Terms Used in National Weather Service Forecasts: Does the Public Understand These Terms? A User's Survey." *Central Region Highlights.* Kansas City: NWS Central Region Headquarters.

Lamont, Michèle, and Laurent Thévenot. 2000. *Rethinking Comparative Cultural Sociology.* New York: Cambridge University Press.

Landow, George P. 2006. *Hypertext 3.0: Critical Theory and New Media in an Era of Globalization.* Baltimore: Johns Hopkins University Press.

Langley, Ann, Henry Mintzberg, Patricia Pitcher, Elizabeth Posada, and Jan Saint-Macary. 1995. "Opening Up Decision Making: The View from the Black Stool." *Organization Science* 6: 260–79.

La Nuez, Danny, and John Jermier. 1994. "Sabotage by Managers and Technocrats: Neglected Patterns of Resistance at Work." In *Resistance and Power in Organizations*, edited by John Jermier, David Knights, and Walter Nord, 219–51. London: Routledge.

Latour, Bruno. 1983. "Give Me a Laboratory and I Will Raise the World." In *Science Observed: Perspectives on the Social Studies of Science*, edited by Karin Knorr Cetina and Michael Mulkay, 141–69. Beverly Hills, CA: Sage.
———. 1987. *Science in Action: How to Follow Scientists and Engineers through Society.* Cambridge, MA: Harvard University Press.
———. 1992. "Where Are the Missing Masses? The Sociology of a Few Mundane Artifacts." In *Shaping Technology/Building Society: Studies in Socio-technical Change*, edited by Wiebe E. Bijker and John Law, 225–58. Cambridge, MA: MIT Press.
———. 1999. *Pandora's Hope: Essays on the Reality of Science Studies.* Cambridge, MA: Harvard University Press.
Latour, Bruno, and Steve Woolgar. (1979) 1986. *Laboratory Life: The Social Construction of Scientific Facts.* Princeton, NJ: Princeton University Press.
Lave, Jean, and Étienne Wenger. 1990. *Situated Learning: Legitimate Peripheral Participation.* Cambridge: Cambridge University Press.
Law, John. 2007. "Pinboards and Books: Learning, Materiality and Juxtaposition." In *Education and Technology: Critical Perspectives, Possible Futures*, edited by David Kritt and Lucien T. Winegar, 125–50. Lanham, MD: Lexington Books.
Law, John, and Annemarie Mol, eds. 2002. *Complexities: Social Studies of Knowledge Practices.* Durham, NC: Duke University Press.
Lawrence, Thomas, Roy Suddaby, and Bernard Leca. 2009. "Introduction: Theorizing and Studying Institutional Work." In *Institutional Work: Actors and Agency in Institutional Studies of Organizations*, edited by Thomas Lawrence, Roy Suddaby, and Bernard Leca. Cambridge: Cambridge University Press.
Lazo, Jeffrey K., Rebecca E. Morss, and Julie L. Demuth. 2009. "300 Billion Served: Sources, Perceptions, Uses, and Values of Weather Forecasts." *Bulletin of the American Meteorological Society* 90: 785–98.
Lenoir, Remi. 2006. "Scientific Habitus: Pierre Bourdieu and the Collective Intellectual." *Theory, Culture and Society* 23: 25–43.
Leonardi, Paul, S. 2011. "When Flexible Routines Meet Flexible Technologies: Affordance, Constraint, and the Imbrication of Human and Material Agencies." *MIS Quarterly* 35: 147–67.
Leschziner, Vanina, and Adam Isaiah Green. 2013. "Thinking about Food and Sex: Deliberate Cognition in the Routine Practices of a Field." *Sociological Theory* 31: 116–44
Lindell, Michael K., and Harold Brooks. 2013. "Workshop on Weather Ready Nation: Science Imperatives for Severe Thunderstorm Research." *Bulletin of the American Meteorological Society* 94: 171–74.
Lindell, Michael K., and Ronald W. Perry. 1992. *Behavioral Foundations of Community Emergency Planning.* Washington, DC: Hemisphere.
Lipshitz, Raanan, Gary Klein, Judith Orasanu, and Eduardo Salas. 2001. "Taking Stock of Naturalistic Decision-making." *Journal of Behavioral Decision-Making* 14: 331–52.

Lipsky, Michael. 1980. *Street-Level Bureaucracy: Dilemmas of the Individual in Public Services*. New York: Russell Sage Foundation.

Lizardo, Omar, and Michael Strand. 2010. "Skills, Toolkits, Contexts and Institutions: Clarifying the Relationship between Different Approaches to Cognition in Cultural Sociology." *Poetics* 38: 204–27.

Lorenzen, Janet A. 2012. "Going Green: The Process of Lifestyle Change." *Sociological Forum* 27: 94–116.

Lynch, Michael. 1985. *Scientific Practice and Ordinary Action: Ethnomethodology and Social Studies of Science*. Cambridge: Cambridge University Press.

Lynch, Peter. 2006. *The Emergence of Numerical Weather Prediction: Richardson's Dream*. Cambridge: Cambridge University Press.

Lyng, Stephen. 1990. "Edgework: A Social Psychological Analysis of Voluntary Risk-Taking." *American Journal of Sociology* 95: 851–86.

MacKenzie, Donald A. 1981. *Statistics in Britain: The Social Construction of Scientific Knowledge*. Edinburgh: Edinburgh University Press.

———. 1990. *Inventing Accuracy: A Historical Sociology of Nuclear Missile Guidance*. Cambridge, MA: MIT Press.

———. 2006. *An Engine, Not a Camera: How Financial Models Shape Markets*. Cambridge, MA: MIT Press.

———. 2009 *Material Markets: How Economic Agents Are Constructed*. Oxford: Oxford University Press.

MacKenzie, Donald A., Fabian Muniesa, and Lica Sieu. 2007. *Do Economists Make Markets? On the Performativity of Economics*. Princeton, NJ: Princeton University Press.

Mackillop, William J. 2008. "Differences in Prognostication between Early and Advanced Cancer." In *Prognosis in Advanced Cancer*, edited by P. Glare and N. Christakis, 13–23. Oxford: Oxford University Press.

Manovich, Lev. 2001. *The Language of New Media*. Cambridge, MA: MIT Press.

Maranta, Allesandro, Michael Guggenheim, Priska Gisler, and Christian Pohl. 2003. "The Reality of Experts and the Imagined Lay Person." *Acta Sociologica* 46: 150–65.

March, James G., and Johan P. Olsen. 1976. *Ambiguity and Choice in Organizations*. Bergen: Universitetsforlaget.

———. 2006. "The Logic of Appropriateness." In *The Oxford Handbook of Public Policy*, edited by Michael Moran, Martin Rein, and Robert E. Goodin, 1–28. Oxford: Oxford University Press.

March, James G., and Herbert A. Simon. 1958. *Organizations*. New York: Wiley.

Marshall, Timothy P., Richard F. Herzog, Scott J. Morrison, and Steven R. Smith. 2002. "Hail Damage Threshold Sizes for Common Roofing and Siding Materials." In *21st Conference on Severe Local Storms, 12–16 August 2002, San Antonio, Texas*, 95–98. Boston: American Meteorological Society.

Martin, John Levi. 2003. "What is Field Theory?" *American Journal of Sociology* 109: 1–49.

———. 2011. *The Explanation of Social Action*. Oxford: Oxford University Press.

Mathis, Nancy. 2007. *Storm Warning: The Story of a Killer Tornado.* New York: Touchstone.

Maximuk, Lynn P. 2003. "Challenges and Rewards of Preparing Gridded Forecasts." Interactive Symposium on AWIPS, American Meteorological Society, Orlando, FL. http://ams.confex.com/ams/pdfpapers/52754.pdf.

McGehan, Barbara. 2002. "Forecasters Learn More about New Forecast Techniques." http://www.oar.noaa.gov/spotlite/archive/spot_gfesuite.html (no longer posted).

Mead, George Herbert. 1932. *The Philosophy of the Present.* Chicago: University of Chicago Press.

Mehan, Hugh. 1993. "Beneath the Skin and between the Ears: A Case Study in the Politics of Representation." In *Understanding Practice: Perspectives on Activity and Context,* edited by Jean Lave, 241–69. Cambridge: Cambridge University Press.

Merleau-Ponty, Maurice. 1962. *Phenomenology of Perception.* London: Routledge.

Merton, Robert, K. 1957. *Social Theory and Social Structure.* Glencoe, IL: Free Press.

Mileti, Dennis, S. 1999. *Disasters by Design: A Reassessment of Natural Hazards in the United States.* Washington, DC: Joseph Henry Press.

Mische, Ann. 2009. "Projects and Possibilities: Researching Futures in Action." *Sociological Forum* 24: 694–704.

Mody, Cyrus M. 2005. "The Sounds of Science: Listening to Laboratory Practice." *Science, Technology, and Human Values* 30: 175–98.

Mol, Annemarie. 2002. *The Body Multiple: Ontology in Medical Practice.* Durham, NC: Duke University Press.

Monmonier, Mark J. 1999. *Air Apparent: How Meteorologists Learned to Map, Predict, and Dramatize Weather.* Chicago: University of Chicago Press.

Morrill, Calvin, Mayer N. Zald, and Hayagreeva Rao. 2003. "Covert Political Conflict in Organizations: Challenges from Below." *Annual Review of Sociology* 30: 391–415.

Morss, Rebecca E., Jeffrey K. Lazo, and Julie L. Demuth. 2010. "Examining the Use of Weather Forecasts in Decision Scenarios: Results from a U.S. Survey with Implications for Uncertainty Communication." *Meteorological Applications* 17: 149–62.

Mukerji, Chandra. 1989. *A Fragile Power: Scientists and the State.* Princeton, NJ: Princeton University Press.

Murphy, Allan H. 1993. "What Is a Good Forecast? An Essay on the Nature of Goodness in Weather Forecasting." *Weather and Forecasting* 8: 281–93.

Murphy, Allan H., and J. C. Curtis. 1985. "Public Interpretation and Understanding of Forecast Terminology: Some Results of a Newspaper Survey in Seattle, Washington." *Bulletin of the American Meteorological Society* 66: 810–19.

Murphy, Allan H., S. Lichtenstein, B. Fischoff, and R. L. Winkler. 1980. "Misinterpretations of Precipitation Probability Forecasts." *Bulletin of the American Meteorological Society* 61: 695–701.

Murphy, Allan H., and Robert L. Winkler. 1974. "Probability Forecasts: A Survey of

National Weather Service Forecasters." *Bulletin of the American Meteorological Society* 55: 1449–52.

———. 1977. "Reliability of Subjective Probability Forecasts of Precipitation and Temperature." *Applied Statistics* 26: 41–47.

———. 1979. "Probabilistic Temperature Forecasts: The Case for an Operational Program." *Bulletin of the American Meteorological Society* 60: 12–19.

———. 1992. "Diagnostic Verification of Probability Forecasts." *International Journal of Forecasting* 7: 435–55.

Myers, Natasha. 2006. "Animating Mechanism: Animations and the Propagation of Affect in the Lively Arts of Protein Modelling." *Science Studies* 19: 6–30.

NOAA (National Oceanic and Atmospheric Administration). 2006. "Policy on Partnerships in the Provision of Environmental Information." http://www.noaa.gov/partnershippolicy/.

NRC (National Research Council). 2003. *Critical Issues in Weather Modification Research*. Washington DC: National Academies Press.

———. 2006. *Completing the Forecast: Characterizing and Communicating Uncertainty for Better Decisions Using Weather and Climate Forecasts*. Washington DC: National Academies Press.

———. 2012. *Weather Services for the Nation: Second to None*. Washington DC: National Academies Press.

NWS (National Weather Service). 2003a. "Weather Forecast Office Operations Vision and Philosophy." http://www.nws.noaa.gov/com/files/opsphil.pdf.

———. 2003b. "Weather Images: Key to Symbols and Abbreviations." http://www.erh.noaa.gov/box/gfe/weathersymbols.html.

———. 2004a. "National Weather Service Instruction 10-506. Digital Data Products/Services Specification." http://www.nws.noaa.gov/ost/ifps_sst/10-506.pdf.

———. 2004b. "Regional Best Collaboration Practices." http://www-md.fsl.noaa.gov/IFPS/2004ws4/RegionalCollab.ppt (no longer posted).

———. 2005a. "National Weather Service Instruction 10-813. Terminal Aerodrome Forecasts." http://www.nws.noaa.gov/wsom/manual/archives/ND319705.HTML.

———. 2005b. "National Weather Service Eastern Region Supplement 01-2004. Winter Storm Verification in Eastern Region." http://www.nws.noaa.gov/directives/010/archive/pd01016001e012004a.pdf.

———. 2005c. "Weather Definition. Predominant Weather (Wx)". http://graphical.weather.gov/definitions/defineWx.html.

———. 2009a. "National Weather Service Instruction 10-13011. Supplementary Observations and Complementary Data Sources and Networks." http://www.weather.gov/directives/sym/pd01013011curr.pdf (no longer posted).

———. 2009b. "National Weather Service Instruction 10-813. Terminal Aerodrome Forecasts." http://www.weather.gov/directives/sym/pd01008013curr.pdf.

———. 2009c. "Why One Inch Hail Criterion?" http://www.weather.gov/oneinchhail/.

———. 2010. "National Weather Service Instruction 10-512. National Severe Weather Products Specification." http://www.weather.gov/directives/sym/pd01005012curr.pdf.

———. 2011a. "National Weather Service Instruction 10-1601. Verification." http://www.nws.noaa.gov/directives/sym/pd01016001curr.pdf.

———. 2011b. "Performance and Awareness Division: Branches." http://www.weather.gov/om/os/pad/padbranches.shtml.

———. 2012a. "Coastal Warning Display Program." http://www.nws.noaa.gov/om/marine/cwd.htm.

———. 2012b. "FY 2013 Budget Highlights." http://www.corporateservices.noaa.gov/nbo/fy13_budget_highlights/NWS_FY13_One_Pager.pdf.

———. 2012c. "National Weather Service Instruction 10-303. Marine and Coastal Services Standards and Guidelines." http://www.nws.noaa.gov/directives/sym/pd01003003curr.pdf.

———. 2012d. "National Weather Service Eastern Region Supplement 02-2003. Winter Weather Watch/Warning/Advisory Procedures and Thresholds." http://www.nws.noaa.gov/directives/sym/pd01005013e022003curr.pdf.

———. 2013a. "National Weather Service Instruction 10-602. Tropical Cyclone Coordination and Emergency Operations." http://www.nws.noaa.gov/directives/sym/pd01006002curr.pdf.

———. 2013b. "National Weather Service Instruction 10-513. WFO Winter Weather Products Specification." http://www.nws.noaa.gov/directives/sym/pd01005013curr.pdf.

———. 2014. "NOAA's National Weather Service: Graphical Forecast." http://graphical.weather.gov/.

NWS Focus. 2004. "Digital Services: Operational Decision Postponed for Further Study" http://www.nws.noaa.gov/com/nwsfocus/fs040504.htm.

Nebeker, Frederik. 1995. *Calculating the Weather: Meteorology in the 20th Century*. San Diego, CA: Academic Press.

Nowotny, Helga. 2005. "The Changing Nature of Public Science." In *The Public Nature of Science under Assault: Politics, Markets, Science, and the Law*, edited by Helga Nowotny, Dominique Pestre, Eberhard Schmidt-Aßmann, Helmut Schulze-Fielitz, and Hans-Heinrich Trute, 1–27. Heidelberg: Springer.

Nowotny, Helga, Peter Scott, and Michael Gibbons. 2001. *Re-Thinking Science: Knowledge and the Public in an Age of Uncertainty*. London: Polity.

Nutt, Paul C, and David C. Wilson. 2010. "Discussion and Implications: Toward Creating a Unified Theory of Decision Making." In *Handbook of Decision Making*, edited by Paul C. Nutt and David C. Wilson, 645–77. New York: Wiley.

Nye, Joseph, Jr. 2008. *Powers to Lead*. New York: Oxford University Press.

Okely, Judith. 2007. "Fieldwork Embodied." *Sociological Review* 55: 65–79.

Oreskes, Naomi. 2003. "The Role of Quantitative Models in Science." In *Models in Ecosystem Science*, edited by Charles D. Canham, Jonathan J. Cole, and William K. Lauenroth, 13–31. Princeton, NJ: Princeton University Press.

Orlikowski, Wanda, J. 2006. "Material Knowing: The Scaffolding of Human Knowledgeability." *European Journal of Information Systems* 15: 522–24.

Orlikowski, Wanda J., and Susan V. Scott. 2008. "Sociomateriality: Challenging the Separation of Technology, Work and Organization." *Annals of the Academy of Management* 2: 433–74.

Orlikowski, Wanda J., and Joanne Yates. 2002. "It's About Time: Temporal Structuring in Organizations." *Organization Science* 13: 684–700.

Orr, Julian E. 1996. *Talking about Machines: An Ethnography of a Modern Job*. Ithaca, NY: Cornell University Press.

Ottinger, Gwen. 2010. "Buckets of Resistance: Standards and the Effectiveness of Citizen Science." *Science, Technology, and Human Values* 35: 244–70.

Overdevest, Christine. 2011. "Towards a More Pragmatic Sociology of Markets." *Theory and Society* 40: 533–52.

Owens, Timothy J., Dawn T. Robinson, and Lynn Smith-Lovin. 2010. "Three Faces of Identity." *Annual Review of Sociology* 36: 477–99.

Perrow, Charles. 1984. *Normal Accidents: Living with High-Risk Technologies*. New York: Basic.

Pew Research Center. 2011. "How People Learn about the Local Community." http://pewinternet.org/~/media/Files/Reports/2011/Pew%20Knight%20Local%20News%20Report%20FINAL.pdf.

———. 2012. "Trends in News Consumption: 1991–2012: In Changing News Landscape, Even Television Is Vulnerable." September 27. http://www.people-press.org/files/legacy-pdf/2012%20News%20Consumption%20Report.pdf.

Pickering, Andrew. 1995. *The Mangle of Practice: Time, Agency and Science*. Chicago: University of Chicago Press.

Pickstone, John V. 1990. "A Profession of Discovery: Physiology in Nineteenth-Century History." *British Journal of the History of Science* 23: 207–16.

Pietruska, Jamie L. 2011. "US Weather Bureau Chief Willis Moore and the Reimagination of Uncertainty in Long-Range Forecasting." *Environment and History* 17: 79–105.

Pielke, Roger A., Jr., and Richard E. Carbone. 2002. "Weather Impacts, Forecasts, and Policy: An Integrated Perspective." *Bulletin of the American Meteorological Society* 83: 393–403.

Pinch, Trevor. 2008. "Technology and Institutions: Living in a Material World." *Theory and Society* 37: 461–83.

Pinch, Trevor, and Wiebe E. Bijker. 1987. "The Social Construction of Facts and Artifacts: Or How the Sociology of Science and the Sociology of Technology Might Benefit Each Other." In *The Social Construction of Technological Systems: New Directions in the Sociology and History of Technology*, edited by Wiebe E. Bijker, Thomas Hughes, and Trevor Pinch, 17–50. Cambridge, MA: MIT Press.

Pinch, Trevor, and Christine Leuenberger. 2006. "Researching Scientific Controversies: The STS Perspective." EASTS Conference "Science, Controversy, and

Democracy," National Taiwan University, Taiwan, August 3–5. http://stspo.ym.edu.tw/easts/abstract.htm.

Pirtle Tarp, Keli. 2001. "Communication in the Distributed Cognition Framework: An Ethnographic Study of a National Weather Service Forecast Office." Unpublished manuscript, University of Oklahoma.

Pixley, Jocelyn. 2009. "Time Orientations and Emotion-Rules in Finance." *Theory and Society* 38: 383–400.

Pliske, Rebecca M., Beth Crandall, and Gary Klein. 2004. "Competence in Weather Forecasting." In *Psychological Investigations of Competent Decision-Making*, edited by J. Shanteau, P. Johnson, and K. Smith, 40–70. Cambridge: Cambridge University Press.

Pollock, Neil. 2005. "When Is a Work-Around? Conflict and Negotiation in Computer Systems Development." *Science, Technology and Human Values* 30: 496–514.

Powell, Walter W., Kelley Packalen, and Kjersten Whittington. 2012. "Organizational and Institutional Genesis: The Emergence of High-Tech Clusters in the Life Sciences." In *The Emergence of Organization and Markets*, edited by J. Padgett and W. Powell, 434–65. Princeton, NJ: Princeton University Press.

Preda, Alex. 1999. "The Turn to Things: Arguments for a Sociological Theory of Things." *Sociological Quarterly* 40: 347–66.

———. 2006. "Socio-Technical Agency in Financial Markets: The Case of the Stock Ticker." *Social Studies of Science* 36: 753–82.

Prentice, Rachel. 2005. "The Anatomy of a Surgical Simulation: The Mutual Articulation of Bodies in and through the Machine." *Social Studies of Science* 35: 837–66.

Racy, J. P. 1998. "How Northeast Indiana and Northwest Ohio Residents Interpret Meteorological Terminology and Services Through NOAA Weather Radio". *NOAA Technical Service Publications*, NWS CR-05.

Ramirez-Sanchez, Saudiel, and Evelyn Pinkerton. 2009. "The Impact of Resource Scarcity on Bonding and Bridging Social Capital: the Case of Fishers' Information-Sharing Networks in Loreto, BCS, Mexico." *Ecology and Society* 14: 22. http://www.ecologyandsociety.org/vol14/iss1/art22/.

Randalls, Samuel. 2010. "Weather Profits: Weather Derivatives and the Commercialization of Meteorology." *Social Studies of Science* 40: 705–30.

Reason, James 1990. *Human Error.* Cambridge: Cambridge University Press.

Rezek, Alan. 2002. "Changing the Operational Paradigm with Interactive Forecast Preparation System: IFPS." Interactive Symposium on AWIPS. American Meteorological Society, Orlando, FL, January 13–17. http://ams.confex.com/ams/pdfpapers/27063.pdf.

Rubinstein, Ariel. 1998. *Modeling Bounded Rationality.* Cambridge, MA: MIT Press.

Ruth, David P. 2002. "Interactive Forecast Preparation: The Future Has Come." In *Preprints: Interactive Symposium on the Advanced Weather Interactive Processing System (AWIPS), 13–17 January 2002, Orlando, Florida*, 20–22. Boston: American Meteorological Society.

Sabel, Charles F. 2005. "A Real Time Revolution in Routines." In *The Firm as a Collaborative Community: Reconstructing Trust in the Knowledge Economy*, edited by C. Heckscher and P. S. Adler: 106–56. Oxford: Oxford University Press.
Sanders, Robert E., and Richard Westergard. 2002. "Empowering the Forecast Consumer: An Investigation of Citizen Need for, and the Technology for Communicating, Process-Centered Weather Information". Final report for the COMET/UCAR Program, Boulder, CO. http://www.comet.ucar.edu/outreach/abstract_final/9893883.htm.
Schaffer, Simon. 1998. "Physics Laboratories and the Victorian Country House." In *Making Space for Science*, edited by Crosbie Smith and Jon Agar, 149–80. London: Macmillan.
Schatzki, Theodor R., Karin Knorr Cetina, and Eike von Savigny. 2001. *The Practice Turn in Contemporary Theory*. London: Routledge.
Schatzki, Theodor R. 1996. *Social Practices: A Wittgensteinian Approach to Human Activity and the Social*. Cambridge: Cambridge University Press.
Schneiderhan, Erik. 2011. "Pragmatism and Empirical Sociology: The Case of Jane Addams and Hull-House, 1889–1895." *Theory and Society* 40: 589–617.
Schön, Donald A. 1983. *The Reflective Practitioner: How Professionals Think in Action*. New York: Basic.
Schütz, Alfred. 1944. "The Stranger: An Essay in Social Psychology." *American Journal of Sociology* 49: 499–507.
——. 1967. *The Phenomenology of the Social World*. Evanston, IL: Northwestern University Press.
——. 1970. *On Phenomenology and Social Relations*. Chicago: University of Chicago Press.
Scott, James C. 1989. "Everyday Forms of Resistance." In *Everyday Forms of Peasant Resistance*, edited by Forrest D. Colburn, 3–33. Armonk, NY: M. E. Sharpe.
——. 1998. *Seeing like a State: How Certain Schemes to Improve the Human Condition Have Failed*. New Haven, CT: Yale University Press.
Secord, Anne. 1994. "Science in the Pub: Artisan Botanists in Early Nineteenth-Century Lancashire." *History of Science* 32: 269–315.
Sewell, William H., Jr. 1992. "A Theory of Structure: Duality, Agency, and Transformation." *American Journal of Sociology* 98: 1–29.
Shanteau, James. 2001. "What Does It Mean When Experts Disagree?" In *Linking Expertise and Naturalistic Decision-Making*, edited by Eduardo Salas and Gary Klein, 229–44. Mahwah, NJ: Lawrence Erlbaum.
Shapin, Steven. 1988. "The House of Experiment." *Isis* 79: 373–404.
——. 1989. "The Invisible Technician." *American Scientist* 77: 554–63.
——. 1994. *A Social History of Truth: Civility and Science in Seventeenth Century England*. Chicago: University of Chicago Press.
Silver, Daniel. 2011. "The Moodiness of Action." *Sociological Theory* 29: 199–222.
Simon, Herbert A. 1947. *Administrative Behavior: A Study of Decision-Making Processes in Administrative Organization*. New York: The MacMillan Company.
——. 1957. *Models of Man: Social and Rational*. New York: Wiley.

———. 1973. "The Structure of Ill Structured Problems." *Artificial Intelligence* 4: 181–201.
Sink, S. A. 1995. "Determining the Public's Understanding of Precipitation Forecasts: Results of a Survey." *National Weather Digest* 19 (3): 9–15.
Smith, Dena T., and Jennifer Hemler. 2014. "Constructing Order: Classification and Diagnosis." In *Social Issues in Diagnosis: An Introduction for Students and Clinicians*, edited by Annemarie Goldstein Jutel and Kevin Dew, 15–32. Baltimore: Johns Hopkins University Press.
Snellman, Leonard W. 1982. "Impact of AFOS on Operational Forecasting". In *Weather Forecasting and Analysis: 9th Conference, Seattle, June/July 1982, Preprints*, 13–16. Boston: American Meteorological Society.
Snook, Scott A. 2000. *Friendly Fire: The Accidental Shootdown of U.S. Black Hawks over Northern Iraq*. Princeton, NJ: Princeton University Press.
Sorensen, John H. 2000. "Hazard Warning Systems: A Review of 20 Years of Progress." *Natural Hazards Review* 1: 119–25.
Spiegler, David B. 1996. "A History of Private Sector Meteorology." In *Historical Essays on Meteorology, 1919–1995*, edited by James Rodger Fleming, 417–41. Boston: American Meteorological Society.
Stanski, Henry R., Laurence J. Wilson, and William R. Burrows. 1989. *Survey of Common Verification Methods in Meteorology*. 2nd ed. World Weather Watch Technical Report 8. Geneva: World Meteorological Organization.
Star, Susan L. 2010. "This Is Not a Boundary Object: Reflections on the Origin of a Concept." *Science, Technology, and Human Values* 35: 601–17.
Stark, David. 2009. *A Sense of Dissonance: Account of Worth in Economic Life*. Princeton, NJ: Princeton University Press.
Stewart, Thomas R., Roger Pielke Jr., and Radhika Nath. 2004: "Understanding User Decision-Making and the Value of Improved Precipitation Forecasts: Lessons from a Case Study." *Bulletin of the American Meteorological Society* 85: 223–35.
Stewart, Thomas R., Paul J. Roebber, and Lance F. Bosart. 1997. "The Importance of the Task in Analyzing Expert Performance." *Organizational Behavior and Human Decision Processes* 69: 205–19.
Stilgoe, Jack. 2007. "The (Co-)Production of Public Uncertainty: UK Scientific Advice on Mobile Phone Health Risks." *Public Understanding of Science* 16: 45–61.
Street, Alice. 2011. "Artefacts of Not-Knowing: The Medical Record, Diagnosis and the Production of Uncertainty in Papua New Guinean Biomedicine." *Social Studies of Science* 41: 1–20.
Suchman, Lucy 1987. *Plans and Situated Actions: The Problem of Human-Machine Communication*. Cambridge: Cambridge University Press.
Swidler, Ann. 1986. "Culture in Action: Symbols and Strategies." *American Sociological Review* 51: 273–86.
———. 2001. *Talk of Love: How Culture Matters*. Chicago: University of Chicago Press.
Tavory, Iddo, and Nina Eliasoph. 2013. "Coordinating Futures: Towards a Theory of Anticipation." *American Journal of Sociology* 118: 908–42.

Timmermans, Stefan, and Marc Berg. 2003. *The Gold Standard: The Challenge of Evidence-Based Medicine and Standardization in Health Care*. Philadelphia: Temple University Press.

Timmermans, Stefan, and Steven Epstein. 2010. "A World of Standards but Not a Standard World: Toward a Sociology of Standards and Standardization." *Annual Review of Sociology* 36: 69–89.

Thomas, William I. 1928. *The Child in America: Behavior Problems and Programs*. New York: Knopf.

Thorlindsson, Thorolfur. 1994. "Skipper Science: A Note on the Epistemology of Practice and the Nature of Expertise." *Sociological Quarterly* 35: 329–45.

Thornes, John E., and David E. Stephenson. 2001. "How to Judge the Quality and Value of Weather Forecast Products." *Meteorological Applications* 8: 307–14.

Thornton, Patricia. H., William Ocasio, and Michael Lounsbury. 2012. *The Institutional Logics Perspective: A New Approach to Culture, Structure, and Process*. Oxford: Oxford University Press.

Tversky, Amos, and Daniel Kahneman. 1974. "Judgment under Uncertainty: Heuristics and Biases." *Science* 185: 1124–31.

Tversky, Barbara. 1993. "Cognitive Maps, Cognitive Collages, and Spatial Mental Models." In *Spatial Information Theory: Theoretical Basis for GIS*, edited by A. Frank and I. Campari, 14–24. Heidelberg: Springer.

Twain, Mark. 1892. *The American Claimant*. New York: Webster.

U.S. Department of Commerce, Office of Inspector General. 1998. "NWS's Verification System for Severe and Hazardous Weather Forecasting Needs Modernization." http://www.oig.doc.gov/OIGPublications/NOAA-IPE-09255-01-1998.pdf.

Van Bussum, Larry. 1999. "A Composite Look at Weather Surveys: Using Several Weather Surveys to Get an Estimate of Public Opinion." NWSO Western Region Technical Attachment no. 99-20, September 28. http://www.wrh.noaa.gov/wrh/99TAs/9920/index.html.

Van Overwalle, Frank, and Marie Vandekerckhove. 2013. "Implicit and Explicit Social Mentalizing: Dual Processes Driven by a Shared Neural Network." *Frontiers in Human Neuroscience* 7: 560.

Vaughan, Diane W. 1996. The *Challenger Launch Decision: Risky Technology, Culture, and Deviance at NASA*. Chicago: University of Chicago Press.

———. 1999a. "The Dark Side of Organizations: Mistake, Misconduct and Disaster." *Annual Review of Sociology* 25: 271–305.

———. 1999b. "The Role of the Organization in the Production of Techno-Scientific Knowledge." *Social Studies of Science* 29: 913–43.

Verghese, Abraham, Erika Brady, Carl Constanzo Kapur, and Ralph I. Horwitz. 2011. "The Bedside Evaluation: Ritual and Reason." *Annals of Internal Medicine* 155: 550–53.

Wagner-Pacifici, Robin. 2000. *Theorizing the Standoff: Contingency in Action*. Cambridge: Cambridge University Press.

Wakefield, Joseph S. 1993. "The AWIPS Forecast Preparation System." http://fxa

.noaa.gov/publications/MD/eft/publications/papers/yellowbook
/introduction.html.
Watling, Richard. 2004. "ERH Max/Min Temp Study." Internal memorandum, NWS Eastern Region Headquarters.
Weeks, John R. 2004. *Unpopular Culture: The Ritual of Complaint in a British Bank.* Chicago: University of Chicago Press.
Weick, Karl E. 1993. "The Collapse of Sensemaking in Organizations: The Mann Gulch Disaster." *Administrative Science Quarterly* 38: 628–52.
———. 1995. *Sensemaking in Organizations.* Thousand Oaks, CA: Sage.
———. 1999. "The Aesthetic of Imperfection in Orchestras and Organizations." In *Readings in Organization Science: Organizational Change in a Changing Context,* edited by M. P. Cunha and C. A. Marques, 541–63. Lisbon: ISPA.
Weick, Karl E., and Karlene H. Roberts. 1993. "Collective Mind in Organizations: Heedful Interrelating on Flight Decks." *Administrative Science Quarterly* 38: 357–81.
Weick, Karl E., and Kathleen M. Sutcliffe. 2003. "Hospitals as Cultures of Entrapment: A Re-Analysis of the Bristol Royal Infirmary." *California Management Review* 45: 73–8.
Weick, Karl E., Kathleen M. Sutcliffe, and David Obstfeld. 1999. "Organizing for High Reliability: Processes of Collective Mindfulness." In *Research in Organizational Behavior,* edited by R. S. Sutton and B. M. Staw, 81–123. Stanford, CA: Jai Press.
Weiner, Stephen S. 1976. "Participation, Deadlines, and Choice." In *Ambiguity and Choice in Organizations,* edited by James G. March and Johan P. Olsen, 225–50. Oslo: Universitetsforlaget.
Weingart, Peter. 1998. "Science and the Media." *Research Policy* 27: 869–79.
Wenger, Étienne. 1998. *Communities of Practice: Learning, Meaning, and Identity.* Cambridge: Cambridge University Press.
Westergard, Richard, and Robert E. Sanders. 2000. "Beyond IFPS, Empowering Weather Information Consumers." Paper presented at the Third Northeast Regional Operational Workshop, Albany, NY, November 6–7. http://cstar.cestm.albany.edu/nrow/nrow2001.htm.
White, Robert M. 2006. "The Making of NOAA 1963–2005." *History of Meteorology* 3: 55–64.
Whitford, Josh. 2002. "Pragmatism and the Untenable Dualism of Means and Ends: Why Rational Choice Theory Does Not Deserve Paradigmatic Privilege." *Theory and Society* 31: 325–63.
Whitnah, Donald R. 1961. *A History of the United States Weather Bureau.* Urbana: University of Illinois Press.
Woolgar, Steve. 1991. "Configuring the User: The Case of Usability Trials." In *A Sociology of Monsters: Essays on Power, Technology and Domination,* edited by John Law, 57–99. London: Routledge.
Wynne, Brian. 1991. "Knowledges in Context." *Science, Technology, and Human Values* 16: 111–21.

Zaloom, Caitlin. 2004. "The Productive Life of Risk." *Cultural Anthropology* 19: 365–91.

Zehr, Stephen C. 2000. "Public Representations of Scientific Uncertainty about Global Climate Change." *Public Understanding of Science* 9: 85–103.

Zerubavel, Eviatar. 1981. *Hidden Rhythms: Schedules and Calendars in Social Life*. Chicago: University of Chicago Press.

———. 1985. *The Seven-Day Circle: The History and Meaning of the Week*. New York: Free Press.

Zhu, Yuejian, Zoltan Toth, Richard Wobus, Dabid Richardson, and Kenneth Mylne. 2002. "The Economic Value of Ensemble-Based Forecasts." *Bulletin of American Meteorological Society* 83: 73–83.

Ziman, John. 1991. "Public Understanding of Science." *Science, Technology, and Human Values* 16: 99–105.

Zuboff, Shoshana. 1988. *In the Age of the Smart Machine: The Future of Work and Power*. New York: Basic.

Index

Abbott, Andrew, 19, 53, 203
accountability, 115, 133
accuracy: collaboration and, 49, 82; commercial fishers' opinion of NWS, 183, 189–92; consistency and, 123; consumers' need for, 175, 190, 194; forecasting skill, 158; forecast quality as, 226n10; model performance vs., 86–87; of NWS predictions, 28, 115, 137, 166; objectivity and, 53; timeliness and, 20, 115, 120–21, 155; verification vs., 116–17, 118, 127–28, 150–51, 231–32n3; warnings and, 155. *See also* forecast goodness; overforecasting; underforecasting
ACSI (American Customer Satisfaction Index), 171, 234n3
adaptive approach, 109, 111, 117, 213
administrative man, 6, 7, 8
Advanced Weather Interactive Processing System. *See* AWIPS
advisories, 119, 125–26, 167
aesthesis, 72, 74, 78, 97, 146, 157, 206, 209, 215
AFOS (Automation of Field Operations and Services), 34
agency. *See* structure and agency
agricultural weather program, 31
Agriculture Department, 28
airport forecasts, 96–97, 99, 101, 115, 148, 175
alerts, 30, 59, 230n7
algorithms, 46, 51. *See also* numerical weather prediction models
Allison, Graham, 223n5
almanacs, 235n7
ambient weather, 93–94, 101, 102, 104, 110
ambiguity, 3, 21, 50, 92, 111, 146
ambivalence, 128, 147, 150
American Customer Satisfaction Index (ACSI), 171, 234n3
American Meteorological Society, 224n2
American pragmatism, 11, 230n3. *See also* Dewey, John; pragmatist theory of action
anchoring heuristic, 71
anticipation, 89, 119, 139, 143, 145, 153, 161, 164, 182, 212. *See also* prospection
apprenticeship, 4, 62, 69, 72, 110, 209
Area Forecast Discussion, 90–91, 122, 126, 131, 154
Aristotle, 26, 72
Army Signal Corps, 28, 64
articulation work, 98
atmospheric indeterminacy, 21, 92–111; art and science of forecasting, 109–11; forecasting place and, 98–104; ground truth and, 93–98; total observation collage and, 104–9
audiences, 19, 103, 167–73, 176, 188, 193, 203. *See also* consumers
Australia, national weather service in, 32
Automated Surface Observation System, 33, 53, 224n5

261

INDEX

automated weather stations, 29, 33, 74, 94, 96–97, 117, 144, 149–50, 229n1, 232n4
Automation of Field Operations and Services (AFOS), 34
aviation forecasts, 174–75
Aviation/MRF model, 75
AWIPS (Advanced Weather Interactive Processing System), 34, 38, 76, 225n7

Beckert, Jens, 213
behavioral economics, 6, 8, 14, 15
Belvedere office, 65
Berg, Marc, 136
Berner, Boel, 227n6
bias, 71–72, 118, 146, 156, 206. See also *aesthesis*
Black-Scholes model, 213
boats, 177–78, 180, 183–87, 191. See also commercial fishers; marine forecasts
boundary work, 48, 54. See also problem of expertise
bounded rationality, 3, 5–6, 71, 202
Bourdieu, Pierre, 12, 64, 199, 230n3
bow echo radar signature, 143
Braque, Georges, 105
bricolage, 98. See also disciplined improvisation; information bricolage
British Meteorological Office, 51
broadcast meteorologists, 18, 19, 30, 90, 127, 224n2, 227n7
busted forecasts, 118–19, 121–35, 149, 188
butterfly effect, 2

Callon, Michel, 35, 215
Canada, national weather service in, 32, 75
carpet bombing, 46–49, 88
Case Study Library (NWS), 129
Cassell, Eric, 207, 210
causality, 126, 134
chaos theory, 52
chat room discussion, 24, 82–85, 103, 108, 229n14, 229n17, 234n4
Chicago office, 64–65, 87, 94, 116, 228nn9–10
Christakis, Nicholas, 210, 211
chronos, 152–53, 155
Cincinnati Observatory, 28
climate forecasts and modeling, 28, 152–53, 192, 203
Climate Prediction Center, 153, 203
clouds, 40–41, 44, 46, 99, 141, 187, 229n1, 230n6

coastal marine forecast, 185, 236n16. See also marine forecasts
Codtown, 178–82, 190, 191, 236n14
cognition: *aesthesis* and, 72, 74, 146, 157, 206, 209; augmented, 96–97, 110, 162; distributed, 60, 145; dual-process theory of, 156–57, 160, 164, 233n7; externalization of, 16, 59–60, 70, 96; habituation and, 12, 96, 217; materiality of, 11, 21, 74; perception and, 70, 74, 99, 102, 104; screenwork and, 21, 67, 74, 79, 106–7, 111, 213; temporal dimensions of, 22, 138–64; visualizations and, 38, 72, 108
cognitive constraints, 3, 4, 5, 70, 146, 202
cognitive heuristics, 5, 9, 20, 22, 65, 71, 143, 156–57, 164, 197–98, 200–201, 204
cognitive labor, 21, 206. See also cognitive strain
cognitive psychology, 6, 8, 9, 20, 71, 143, 156, 202
cognitive strain, 160, 163, 200, 205, 212
cognitive task analysis, 14
coherence, 11, 20, 98, 105–8, 110, 120, 158–59, 161, 208
collaboration, 60, 78, 80–84, 85, 120, 145, 229nn15–16
collage: defined, 21; heuristic of, 104–9, 111, 138, 161, 208, 216, 231n10; information bricolage and, 106, 110, 208; as screenwork, 21, 106–9, 111
collective mindfulness, 82
comma-shaped radar signature, 143
commercial fishers, 176–96; opinions on NWS, 183, 189–92; practical knowledge of, 186–88, 193; risk management by, 135, 166, 182–84; weather data analysis by, 184–89
communities of practice, 4, 10, 27, 48, 65, 135, 208
computer-generated forecasts. See model-generated forecasts
computer models, 1, 29, 87, 129, 142, 154, 161, 162. See also model-generated forecasts; numerical weather prediction models
controlled chaos, 111, 145, 207, 212
consistency, 42, 49, 81, 82–83, 84, 108, 120, 121, 123, 126, 129, 155, 158, 159, 160, 233n12
consumers: commercial fishers as, 176–84; interactions with, 171–76; NWS distinctions among, 172; NWS functional

262

constructs of, 169, 170, 193, 195, 196; NWS publics, 167–71, 192–96; opinions on NWS, 189–92; weather data analysis by, 44, 184–89; of weather forecasting, 165–96
controversy, 18, 26, 33, 39, 48, 49, 96, 117
Convective Outlooks, 229n13
convective season, 140, 141, 142
Cooperative Weather Observer Program, 229n2
counterperformativity, 215
craftwork, 4, 8, 12–13, 85, 107–9, 110, 231n10
credibility, 19, 25, 49–52, 81, 86, 88, 105, 135, 167, 174, 175, 195
Cuban missile crisis, 146
cultural dopes, 89. *See also* judgmental dopes
cultural repertoires, 9, 12, 20, 201–2, 237n1, 238n2
culture: of apprenticeship, 62, 72; of commercial fishing, 177; of disciplined improvisation, 12, 15, 21, 96, 97–98, 108, 111, 230n3; epistemic, 27, 38, 56, 97, 102, 203, 206; of hazardous weather forecasting, 113; materiality of, 11, 21, 74; of NWS, 18, 29, 48, 55; of NWS forecast offices, 55, 56, 64–65, 149; as tool kit, 12, 70, 201–2
cumulus clouds, 141
Curtis, Rick, 174–75
customers. *See* consumers

Daston, Lorraine, 52, 53, 54
data analysis. *See* weather data analysis
decision making: in action, 3–14; aided, 59, 110, 162; creativity in, 7, 12, 20, 89, 97, 201; deep uncertainty and, 1–23, 215; dual process theory of, 156; emergency, 140–52; emplacement of, 55, 65, 104; end-means continuum of, 13, 211; extended alert, 152–57; fast and frugal, 143–44, 156–57, 164, 204; in finance, 212–16; as habitual action, 10–13, 201, 206; institutionalized environment of, 5–7; long-term, 157–61; in medicine, 207–11; microcontext of, 13–14; as practical action, 10–13; as process, 3–4, 9, 25, 106, 124, 133, 197–99; prospective nature of, 9, 113, 215, 217; risk and, 205–6; short-term, 161–64; sociology of, 3, 197–217; sociomateriality of, 9, 11, 18, 216; in summer weather forecasting, 140–52; systems approach to, 8; temporal regimes of, 22, 138–64; time and, 202–5; weather forecasting as, 14–19
decision support systems, 59. *See also* decision making: aided
deep uncertainty, 2–3, 14, 32, 138, 193, 212–13, 216. *See also* uncertainty
deficit model, 173
deliberation, 78–84, 147, 204, 211
deterministic forecasts, 47, 193
Dewey, John, 11, 12, 13, 16, 67, 201, 211
diagnosis, 13–14, 15, 67, 74, 77, 107, 210, 211. *See also* weather data analysis
Digital Forecast Database: forecast goodness and, 42; gridded forecasts in, 85–86, 90, 160, 225n6, 231n1; implementation of, 35–38, 44, 46–49, 50, 168, 174; interoffice collaboration and, 49, 80, 81, 82; near-term forecasting and, 161; NWS commitment to, 195–96; public opinion on, 195; updating forecasts in, 85; visual aesthetic of, 159
disaster preparedness, 166–67
disaster studies, 8, 223nn6–7
disciplined improvisation, 15, 21, 97–98, 108, 111, 230n3
discretionary power, 7, 32, 52, 78, 88, 98, 130, 214
distributed cognition, 6, 60, 95–96, 145, 208, 213
duality of error, 135–37. *See also* overforecasting; performance standards; underforecasting
Dual-Polarization Doppler Radar, 33. *See also* radar
dual-process theory, 156–57, 160, 164, 233n7

early warning weather system, 119–20
Eastern Region Headquarters, 63, 64, 81, 128
ECMWF. *See* European Center for Medium-Range Weather Forecasts (ECMWF)
ecological rationality, 203. *See also* local rationality
economic forecasting, 212–16
electronics technicians, 61–62, 227n4
emergency decision making, 139, 140–52
emotional labor, 40, 79, 146, 147, 206, 209, 212
empirical context of action, 5, 9, 13–14, 20, 198, 200–201, 204, 237n1
emplacement, 55, 65, 104

263

INDEX

ends-in-view, 71
ends-means continuum of reasoning, 11, 13–14, 15, 67, 71, 73, 77, 198, 211, 213
environmental impatience, 146, 152, 153–54, 203
Environmental Modeling Center, 29
environmental patience, 154, 204
Environmental Science Services Association, 28
epistemic asymmetry, 169, 173. *See also* problem of expertise
epistemic culture, 27, 38, 56, 97, 102, 203, 206
equipment, 9, 16, 26, 56, 61, 69, 201, 204, 213
European Center for Medium-Range Weather Forecasts (ECMWF), 76, 77, 107
Excessive Rainfall Potential Outlook, 228n13
expertise: acquisition process, 4–5, 217; asymmetric information and, 169; of commercial fishers, 188; in communities of thought, 10; context dependence of, 4, 53, 167, 238n2; decision-making standards and, 26, 50; deficit model and, 173; embodied nature of, 37; IFPS and, 54; interactional, 148; intuition and, 96, 143, 146, 156; lay vs. professional, 168–69, 173; microcontext of action and, 13–14; normative definitions of, 4–5; as practical knowledge, 193; problem of, 169, 192–93; screenwork and, 21, 67, 74, 79, 106–7, 111, 213; short-term forecasting and, 163; situated nature of, 4, 48
extended alerts, 139–40, 152–57
extended cognition, 60. *See also* decision making: aided; distributed cognition
Extended Forecast Discussion, 228n13
externalized cognition, 5, 6, 10, 13, 59–60, 107, 157, 205, 215, 217

failure, 8, 104, 110, 188, 204. *See also* busted forecasts
false negatives, 135
false positives, 135
Faraj, Samer, 208
fast and frugal heuristics, 143–44, 156–57, 164, 204. *See also* cognitive heuristics
fatefulness, 152, 163, 205–6, 208, 212, 238n2
Federal Aviation Administration, 224n5, 235n5
Federal Signal Public Safety Survey, 166
Federal Tort Claims Act, 98

field offices, 29, 35, 55, 57. *See also* interoffice collaboration
field science, 99, 102, 105. *See also* laboratory science
finance industry, decision making in, 212–16
Fine, Gary Alan, 18, 29, 31, 38, 55–56, 64–66, 81, 86, 87, 98, 116, 120, 139, 147, 149–50, 176, 228nn9–10, 233n5
fire weather program, 31
fishable weather, 182–84
fishing industry. *See* commercial fishers
fishing regulations, 181, 182, 183, 184, 187, 191
Fligstein, Neil, 223n7
flip-flopping, 86, 129, 154, 158
Flowerland office, 65, 228n10
flurries, 131, 132–33, 134
fog, 114, 131
forecast funnel, 74, 75, 79
forecast goodness: defined, 40–42; IFPS implementation and, 46, 47; NWS administration views on, 86–87
forecast grids, 35, 76, 78, 80, 81, 84, 85, 90, 200. *See also* gridded forecasts
forecasting. *See* weather forecasting
forecasting inertia, 86
forecasting skill, 68, 102, 116, 118, 157–58, 163, 176
forecasting tools. *See* tools
forecast models. *See* numerical weather prediction models
forecast users. *See* consumers
forecast verification. *See* verification
foresight, 67, 211. *See also* anticipation; prospection
Fort, Tom, 165
frost, 114
fundamental uncertainty. *See* deep uncertainty

Galison, Peter, 52, 53, 54
gender work, 62–63, 227n7. *See also* women forecasters
General Forecasters, 62, 72
gestalt-building, 99, 102, 105, 106, 108, 111, 161, 208
GFS. *See* Global Forecast System (GFS)
Gibson, David, 146, 223n5
Giddens, Anthony, 205, 206, 212
Gieryn, Thomas, 19
Gigerenzer, Gerd, 143, 203

Global Forecast System (GFS), 75–76, 79, 83, 87, 122, 159–60, 200
Goffman, Erving, 78, 163, 205
goodness. *See* forecast goodness
Government Paperwork Reduction Act, 170
graphical forecast, 18, 26, 34–35, 38, 42, 50, 54
Graphical Forecast Editor, 34, 60, 76, 78, 80, 88, 107, 162
Great Britain, national weather service in, 32, 75
gridded forecasts, 44–46, 75–76, 84–85, 90, 158
groundfish industry, 179–80, 181, 236n14
ground truth, 3, 93–98, 101, 118, 216
guidance. *See* model guidance; observation data and guidance

habitual decision making, 5, 10–13, 89, 201–2, 205, 206, 208
habituation, 12, 96, 201, 217
hail, 114, 115, 144, 148, 151, 233n6
Hammond, Kenneth, 136
ham radio operator, 145, 147–48
hazardous weather forecasting, 112–37; accuracy of, 115–18; atmospheric indeterminacy and, 21, 103; busted forecasts, 121–35; duality of error in, 135–37; meteorological deliberation and, 80; as primary responsibility, 66; snow events, 121–35; terminology of, 113–15; verification of, 115–18; warning communication system for, 118–21
Hazardous Weather Outlook, 77, 119, 120, 144
Hazardous Weather Warning, 119, 134, 201
Hazardous Weather Watch, 119
hazards mitigation, 166–67
Heidegger, Martin, 231n9
heuristics. *See* cognitive heuristics
high-impact events, 130. *See also* hazardous weather forecasting
Hippocrates, 233n1
HMTs. *See* Hydrometeorological Technicians (HMTs)
hodograph, 142–43
hook echo radar signature, 143
human-nonhuman relations, 9. *See also* practice: sociotechnical nature of; sociomateriality
hurricanes, 92, 112–13, 120, 186, 192, 203. *See also* National Hurricane Center
Hydrometeorological Prediction Center, 29, 153
Hydrometeorological Technicians (HMTs), 59, 62, 80, 227n4

identity narratives, 39–40, 45, 48, 64, 176
idiocultures, 55, 56, 64–65, 149
IFPS. *See* Interactive Forecast Preparation System (IFPS)
impression management, 32, 128, 129, 133
information: bricolage, 98, 106, 110, 208, 230n3; constraints, 3, 195, 202; equivocality, 9; heterogeneity of, 94, 98, 106, 109, 111; overload, 70, 75, 76
information omnivore, 21, 92, 95, 98, 109
infrastructural globalism, 27
infrastructural inversion, 48
Ingold, Tim, 104, 230n7
institutionalized decision-making environment, 5–7, 9, 20, 25–26, 198, 201, 205, 212
institutionalized risk environments, 205, 206, 212
institutional logics: scholarship on, 6–7; in weather forecasting, 32, 51, 56, 88, 131
interaction order, 55, 65
Interactive Forecast Preparation System (IFPS): implementation of, 34–39, 43–51, 55, 66, 90; multismoothing in, 84; objectivity and, 54; resistance to, 44–50. *See also* Graphical Forecast Editor
Interns, 62, 80
interoffice collaboration, 49, 60, 62, 80–83, 122, 181, 217, 229n15
interpretive flexibility of technology, 36
Intersite Coordination Database, 82, 83
interventionism, 136, 210, 211, 215
intuition, 96, 143, 146, 156

James, William, 11
Jermier, John, 45
Johnson, Philander Chase, 1
journeymen, 62. *See also* Junior Forecasters
judgment, 2–6, 10, 25, 52–54, 71, 73, 74, 77, 89, 96, 130, 141, 143, 152, 157, 186, 187, 194, 202, 206, 207, 210
judgmental dopes, 13, 136, 215
Junior Forecasters, 62, 72

kairos, 141, 152–53, 155
Katz, Jack, 14

kidney bean radar signature, 143
Klein, Gary A., 203
Knorr Cetina, Karin, 10, 12, 27, 75, 79, 102
knowledge infrastructures, 15–16, 104, 213, 215, 216

laboratory science, 4, 13, 93, 98–99, 101–2, 105, 109, 207
Langley, Ann, 7
LaNuez, Danny, 45
Latour, Bruno, 10, 32, 48, 74, 104–5, 109, 207
LeFebvre, Tom, 37
legacy NWS products, 35, 90
legitimacy, 48, 93, 98, 113, 129, 135, 167, 169, 195
lightning, 114–15
liminal epistemic space, 109, 207. *See also* field science; laboratory science
local embeddedness, 9, 53. *See also* emplacement
local rationality, 25, 48, 84, 149, 151, 152, 203, 207, 212–13. *See also* practical action
long-term decision making, 139–40, 200, 214
long-term forecasting, 59, 60, 74, 79, 100, 107, 125, 132, 157, 158, 160, 164
Lorenz, Edward, 52

MacKenzie, Donald, 213
managing risk. *See* risk and risk management
Manovich, Lev, 108
Maranta, Allesandro, 169
March, James, 5
marine accidents, 184, 191
marine forecasts, 19, 56, 80, 176, 184, 193. *See also* commercial fishers
Martin, John Levi, 12
materiality: of action, 11, 13, 20; of cognition, 11, 21, 59, 74; of culture, 12, 201–2; of decision making, 6, 13, 74, 85, 88, 89, 198, 213; of practice, 10, 11; of screenwork, 106, 108; of weather, 15, 21, 74, 100, 103–4, 114, 115, 231n1
mediatization of weather, 166
medicine, decision making in, 207–11
Medium-Range Weather Forecasts, 76
Mellstrom, Ulf, 227n6
Merleau-Ponty, Maurice, 99, 231n9
Merton, Robert, 214
mesoscale, 74, 141
Meteorologica (Aristotle), 26

meteorological *aesthesis*, 72, 74, 146, 157, 206, 209, 215
meteorological deliberation. *See* deliberation
meteorological risk. *See* risk and risk management
meteorologists: broadcast, 18, 19, 90, 224n2; data sources for, 95–96; General Forecasters, 62, 72; Interns, 62, 80; obsession with weather, 39–43; operational, 40; Senior Forecasters, 62, 72, 80. *See also* women forecasters
métis, 193
Metrocity, 57, 107–8, 115, 121, 123–28, 131, 132
microclimate, 57, 102, 114, 232n3
microcontext of action. *See* empirical context of action
microscale, 74
midnight shift, 61, 62, 68, 87, 124–26, 128–29, 131–32
missed forecasts. *See* busted forecasts
MM5 (Fifth-Generation Penn State/NCAR Mesoscale Model), 76
model-generated forecasts, 72, 77, 86, 87, 93, 111
model guidance, 24, 52, 54, 78, 83, 84, 87, 117, 129, 144, 159, 162–63
Model Output Statistics (MOS), 88, 159–60, 200
models. *See* numerical weather prediction models; *and specific models by name*
Mol, Annemarie, 208
multismoothing, 84
mutual shaping of technology and society, 5, 36, 74, 215. *See also* sociomateriality

narrative arc of weather forecasts, 159–60
National Centers of Environmental Prediction, 29, 77, 83
National Digital Forecast Database. *See* Digital Forecast Database
National Hurricane Center, 29, 61, 120, 121
National Marine Fisheries Service, 236n17
National Oceanic and Atmospheric Administration (NOAA), 28, 31, 58, 185, 186, 189, 236n17
National Research Council, 51, 228n11, 232n5
National Weather Association, 40
National Weather Service (NWS), 15–22; Case Study Library, 129; Directives,

96, 98, 130–31, 139, 167, 175; duality of error in, 135–37; forecasting routine, 34, 35, 38, 56, 67; hazardous weather forecasting as primary responsibility, 66, 113; IFPS implementation by, 43–51; interactions with broadcast meteorologists by, 18, 19, 90, 224n2; interactions with customers by, 171–76; interactions with reporters by, 133; meteorologists in, 39–43; mission of, 28; modernization of, 33–39, 230n4; objectivity at, 51–54; opinion surveys conducted by, 170–71; organizational structure of, 29, 32, 49, 58–62, 167; origins of, 20–21, 26–33; public opinion on, 189–92; publics of, 167–71, 192–96; relationship with private weather industry, 15, 18, 30, 31, 36, 50, 176; as study focus, 15. *See also specific offices*

Nautical Almanac, 235n7

near-term decision making, 139–40, 161–64

Neborough office, 20–21, 55–91; atmospheric indeterminacy and, 92–111; briefings by outgoing forecaster, 67–71; building tour, 58–62; Chicago office vs., 64–65, 87, 94, 116, 228nn9–10; culture of, 62–65; demographics of, 62; emergency decision making in, 140–52; employees of, 62–65; hazardous weather forecasting in, 112, 121–35; IFPS implementation in, 43–51; location of, 57–58; meteorological deliberation in, 78–84; as study environment, 15; summer weather forecasting in, 140–52; weather data analysis in, 71–78; weather forecast production in, 65–91; winter weather forecasting in, 121–35. *See also* National Weather Service (NWS)

neighboring offices. *See* interoffice collaboration

New Zealand, national weather service in, 32

NGM (Nested Grid Model), 76

NOAA. *See* National Oceanic and Atmospheric Administration (NOAA)

North American Mesoscale (NAM), 75–76, 83, 107

nowcasting, 93, 104, 119, 124, 125

Nowotny, Helga, 176

numerical weather prediction models: atmospheric indeterminacy and, 52; development of, 27; forecast goodness and, 87–88; improvements to, 3; initialization conditions for, 2; as standard to beat, 86–87; weather data analysis and, 16, 74–78, 107. *See also specific models by name*

NWS. *See* National Weather Service (NWS)

NWS forecast offices. *See* Belvedere office; Chicago office; Flowerland office; Neborough office

NWS warning system, 25, 28, 57, 59, 114, 119–22, 123, 125–30, 135, 141, 148–50, 155, 165–66, 169, 195

object-centered sociality, 79
objectivity of weather forecasting, 51–54
observation data and guidance, 29, 30, 52, 72, 73, 75, 77, 93, 128, 142
Ocean Prediction Center, 29, 61
offshore forecast, 236n16
Old Farmer's Almanac, 235n7
operational meteorologists, 40. *See also* identity narratives; professional identity
operational transition, 20, 38, 43, 45, 51
optimality, 4, 6, 156
organic infrastructures, 48
organizational change. *See* operational transition; standardization
organizational decision making. *See* decision making
organizational disasters. *See* disaster studies
organizational logic. *See* institutional logics; standardization
organizational regulation. *See aesthesis*; apprenticeship; habituation
organizational sensemaking. *See* sense making
outreach events, 172
overforecasting, 22, 126, 129, 135–37

pattern recognition, 72, 73, 142–43, 144, 146
People's Weather Service, 165, 169, 170, 196. *See also* National Weather Service (NWS)
perceptual focus, 67, 73, 74, 126, 151–52
performance standards, 6, 135, 136, 137. *See also* verification
performativity, 215
Pew Research Center, 166
Pickering, Andrew, 36
Pinch, Trevor, 10
place for forecasting, 98–104, 109
Plato, 72
ports, 178, 179, 180, 182

Post-Storm Analysis Team, 118, 128, 129
practical action: decision making as, 10–13, 203; weather forecasting as, 75–76, 84–85, 89, 107, 203, 216
practice: cultural retooling and, 202; habituation and, 12, 96, 217; as intelligent achievement, 12; material and symbolic context of, 20, 198; in routine and nonroutine situations, 10, 13, 15, 20, 22, 66, 94, 124, 139; routinized action and, 5, 10, 12, 20, 160, 199, 201; sociotechnical nature of, 5, 6, 13, 48, 215
pragmatist theory of action, 11, 20, 84, 163, 230n3. *See also* practice
precipitation events, 47, 141–42, 172. *See also* rain; showers; snow
Preliminary Operational Assessment, 151
priming, 66, 95, 111, 139, 143, 204. See also *aesthesis*
private weather industry, 15, 18, 30, 31, 36, 50, 176
probability, 2, 47, 192–93, 212, 226n12
problem of expertise, 98, 169, 173, 176, 192–93
Procedures, 69–70, 74
professional autonomy, 32, 73, 77, 86. *See also* discretionary power
professional identity, 39–41, 48, 49, 51, 165
prognosis, 3, 13–14, 20, 21, 57, 67, 76, 77, 107, 210–11
propertization of public science, 31
prospection, 9, 113, 139, 217
public-domain science, 30, 168, 169, 176
public mistrust, 166, 173, 194
publics, 19, 30, 166, 167–71, 192, 196. *See also* audiences; consumers
public safety, 118–20, 127, 166, 174, 175, 194
public science, 30–32, 176, 224n1
public service, 40, 43–44, 47–49, 117, 120, 144, 165, 193

quantitative forecast models. *See* numerical weather prediction models

radar: availability of, 31; gridded forecasts and, 108; ground truth and, 101; interoffice collaboration and, 174; modernization of, 33; at Neborough office, 57; reflectivity patterns, 143; severe weather forecasting and, 144–45, 149, 151–52; severe weather warning team and, 145, 151; in Situational Awareness Display, 60; summer weather forecasting and, 141, 143, 144; tornado signatures on, 113, 141; value of, 142; weather data analysis and, 2, 16, 74, 94; winter weather forecasting and, 123, 124, 128
radar warning duty, 145, 147
rain, 42, 45, 47, 87, 93–94, 101, 103, 112, 117, 135, 141–42, 155, 165. *See also* showers; thunderstorms
Rapid Update Cycle (RUC) model, 76, 77
rational choice theory, 3–4, 5, 7, 67, 212, 214
rationality: *aesthesis* and, 72, 74, 146, 157, 206, 209; boundedness of, 3, 5–6, 71, 202; decision making and, 3–4, 5, 203, 212; expertise and, 10, 26, 50; institutional logics and, 7, 20, 32, 51, 56, 88, 131; intuition and, 96, 143, 146, 156; local, 25, 84, 149, 151, 152, 203; organizational logic and, 18; situated logic and, 211, 217
rationalizations, 8–9, 136, 207. *See also* risk and risk management
rational man. *See* rational choice theory
reactive decision making, 203
reasoning. *See* cognition; decision making; rationality
recipes for action, 200–201, 237n1. *See also* local rationality; stock of knowledge
reflexivity, 56, 193, 214, 215
regulated improvisation, 230n3
relationality, 4–5, 10, 105, 203
research meteorologists, 40
resistance, 36–38, 45, 49, 168, 193
resourcefulness, 25, 89, 201. *See also* bricolage; local rationality
Richardson, Lewis Fry, 51
risk and risk management: commercial fishers and, 182–83; decision making and, 8–9, 205–6; in finance industry, 212–16; in hazardous weather forecasting, 21–23, 112–37; in long-term forecasting, 160; mitigation of, 194–95; predicting risk, 25; sense making and, 9, 146, 153; in short-term forecasting, 163–64; summer weather forecasting, 140, 143–44; temporal embedding of forecasting and, 139; uncertainty and, 2
Roberts, Karlene H., 81–82
routinized decision making, 5, 10, 12, 20, 89, 160, 199–200, 201

Royal Society of London, 27
RUC (Rapid Update Cycle) model, 76, 77

sabotage by circumvention, 45, 48. *See also* resistance
satellite imagery, 60, 73, 74, 94, 96, 101, 108, 123
satisficing heuristics, 6, 71. *See also* cognitive heuristics
scale, 74, 139. *See also* mesoscale; microscale
scallopers, 180, 181, 184
scenario building, 146
Schütz, Alfred, 9, 153, 200–201
scientism, 52, 97, 109
screenwork, 21, 67, 74, 79, 106–7, 111, 213
self-fulfilling prophecy, 214, 215
Senior Forecasters, 62, 72, 80
sense making, 9, 146, 153
sensible weather, 103, 106, 113, 134, 189
Severe Prediction Center (SPC), 120–21
Severe Thunderstorm Warnings, 233n6
severe weather forecasting, 70, 75, 114, 121, 141–47, 150–51, 166, 191. *See also* hazardous weather forecasting; summer weather forecasting
severe weather spotters. *See* weather spotters
Severe Weather Statements, 114, 148
severe weather warning team, 145, 151
shift briefing, 1, 72, 73. *See also* weather briefings
Short Range Ensemble Forecasting (SREF) model, 77
short-term forecasting, 59–61, 68, 70, 74, 79, 87, 93, 99–100, 107, 125, 132, 140, 147, 157, 159–64
showers, 47, 131, 133, 141, 142. *See also* rain; snow
Simon, Herbert A., 5, 6, 7, 8, 71, 224n12
situational awareness, 21, 72. See also *aesthesis*
Situational Awareness Display, 60
situational rationality. *See* local rationality
Small Craft Advisory, 167
Smithsonian Institution, 27, 28
snow, 94, 107, 112, 114, 119–22, 125–28, 131–34, 154–55; emergency procedures, 126–27; showers, 93, 131, 132, 133; squalls, 22, 93, 131, 132–33, 134. *See also* flurries
social embeddedness, 8–9, 25, 32, 53, 213
social invisibility of technicians, 61
socially distributed expertise, 176

socially marked nature of weather, 103, 115, 134
social media, 234n3
sociomateriality, 6–7, 9, 11, 18, 20, 32, 74, 89, 130, 198, 216
Southwest Airlines, 174–75, 234n5
SPC (Severe Prediction Center), 120–21
Special Weather Statements, 119, 148
spotters. *See* weather spotters
squall lines, 143
SREF (Short Range Ensemble Forecasting) model, 77
stakeholders, 36, 43. *See also* publics
standardization: forecasting models and, 25, 36, 43, 46, 48, 72, 86–87, 88; IFPS implementation and, 48; operational transition, 43–51; overforecasting vs. underforecasting, 135–36; politics of, 35, 46, 88; resistance to, 36, 37, 45, 49, 168, 193; of tools, 27, 33–39; for weather instruments and measurements, 27
stock of knowledge, 7–9, 20, 193, 200, 206, 217
storm coordinator, 145, 147
Storm Prediction Center, 29, 30, 61, 120, 229n13
StormReady Community certification, 172
storms. *See* hazardous weather forecasting; snow; thunderstorms
straight-line winds, 140
stratus clouds, 141
Street, Alice, 208
structure and agency, 12, 89, 199, 201
subjectivity of weather forecasting, 51–54
subjunctive mood, 14, 15, 97, 111, 146, 154
summer weather forecasting, 140–52; decision making in, 22, 120, 156; environmental impatience and, 153; local nature of, 149; storms, 80, 112, 139, 142, 146, 152; verification vs. accuracy of, 150–51
Swidler, Ann, 12, 237n1

teamwork, 49, 59, 80, 145, 147. *See also* interoffice collaboration
technological change, 35, 43–51. *See also* operational transition
technologically mediated interaction order, 65, 213
technological scaffolding, 102
technological script, 36, 54, 88, 233n12
technology adoption, 36, 43–51

269

telegraph, 27
temporal forecasting regimes, 22, 138–64; long-term forecasting, 157–61; severe weather forecasting, 140–52; short-term forecasting, 161–64; winter weather forecasting, 152–57
TEMPOs, 175
text forecasts: authoring of, 41–42, 44, 45, 46; IFPS implementation and, 37, 42–47, 90; interoffice collaboration and, 50, 81; long-term forecasting and, 159; shift from, 18, 26, 38
thunderstorms, 1–2, 42, 90, 114–15, 140, 142, 144, 151, 172, 233n6
timeliness, 20, 115, 118–21, 155, 202–5
Timmermans, Stefan, 136
tools, 9, 12, 25, 26, 31, 34, 52, 60, 78, 90, 102, 107–8, 142, 188, 200, 202, 215, 232n5, 233n12. *See also* equipment
Tornado Alley, 145
tornadoes, 113, 114, 140, 141, 206
total observation, 96–98, 100, 104–9. *See also* coherence; collage; information: bricolage
Total Observation Concept, 96, 97, 98, 104–9
Transportation Department, 106, 132, 134
Tropical Prediction Center, 112
trust, 166–67, 173, 187, 194–95
Tversky, Barbara, 231n10
Twain, Mark, 41
Type I error, 135
Type II error, 135

uncertainty: communication of, 90, 166–67, 195; deep, 2–3, 14, 32, 138, 193; fundamental, 212–13; information bricolage and, 106, 110, 208; provisional coherence and, 105–10
underforecasting, 22, 126, 135–37
United Kingdom, national weather service in, 32, 75

verification: accuracy vs., 116–17, 118, 127–28, 150–51, 231–32n3; credibility and, 19, 25, 49–52, 81, 86, 105; forecaster performance and, 116–17; ground truth and, 93–98, 101, 118; metrics, 10, 13, 15, 20, 22, 66, 94, 124, 139; as numbers game, 87, 117, 150; as organizational game, 116, 150
visual confirmation, 100–101
visualizations, 38, 72, 75, 102, 108, 230n8

visual pattern recognition, 72, 73, 142, 144, 146

Wagner-Pacifici, Robin, 14, 139, 154
War Department, 28
Warning Event Simulator, 128
warnings, 25, 28, 57, 59, 114, 119–22, 123, 125–30, 135, 141, 148–50, 155, 165–66, 169, 195
watches, 68, 119–22, 185
weather balloons, 29, 73, 226n1
weather briefings, 67–71, 73–74
Weather Bureau, 28, 51
Weather Channel, 31, 38, 60, 174, 185, 234n1
weather data analysis, 71–78, 135, 184–89. *See also* diagnosis; prognosis
weather forecasting, 24–54; art and science of, 71, 109–11; audience for, 165–96; as collage, 104–9, 138, 161, 208, 216, 231n10; as decision making, 14–19; digital age of, 34, 39, 41, 44, 109, 111; discretionary power in, 7, 32, 52, 78, 88, 98, 130, 214; embodied nature of, 37, 85, 102–3, 200; meteorologists drawn to field of, 39–43; modernization of, 33–39; in Neborough office, 65–91; NWS and, 26–33; physicality of, 16; practice of, 14, 18, 22, 89, 139, 140, 208, 209, 213, 216; push vs. pull models of, 44; quality of, 41, 54, 82; social dimensions of, 10, 11, 69, 86, 129, 130, 134, 152, 202, 215, 230n5; subjectivity of, 51–54; terminology of, 41, 44, 131–33, 226n11. *See also* hazardous weather forecasting; severe weather forecasting; summer weather forecasting; winter weather forecasting
Weather Research and Forecast (WRF) model, 77
weather spotters, 124, 127, 144
Weick, Karl E., 8, 81–82, 126, 223n8
Weiner, Stephen, 146
Whaletown, 178–82
Wilde, Oscar, 197
wind: commercial fishers and, 182, 184, 187, 189, 190–92; ground truth and, 93, 99; long-term forecasting and, 159; measurements of, 94, 97; short-term forecasting and, 162; summer weather forecasting and, 140, 143; terminology of, 226n11; weather data analysis and, 69, 84; winter weather forecasting and, 133

Winter Outlook, 153
Winter Weather Desk, 153
winter weather forecasting: accuracy of, 190; decision making and, 22; environmental patience and, 154; hazardous weather, 112–14, 121, 125, 127, 131, 134; storms, 108, 122, 129, 130, 139, 152, 154
winter weather warnings, 125, 134
winter weather watches, 120, 122, 155

women forecasters, 62, 63, 225n8, 227nn6–7
work-around, 84, 88, 97
WRF (Weather Research and Forecast) model, 77

Xiao, Yan, 208
X-town, 57–58, 93, 124, 125, 148, 149

yo-yo forecasting, 86